Selected Titles in This Series

733 **Milé Krajčevski,** Tilings of the plane, hyperbolic groups and small cancellation conditions, 2001

732 **Jan O. Kleppe, Juan C. Migliore, Rosa Miró-Roig, Uwe Nagel, and Chris Peterson,** Gorenstein liaison, complete intersection liaison invariants and unobstructedness, 2001

731 **Jesús Bastero, Mario Milman, and Francisco J. Ruiz,** On the connection between weighted norm inequalities, commutators and real interpolation, 2001

730 **Suhyoung Choi,** The decomposition and classification of radiant affine 3-manifolds, 2001

729 **Michael Grosser, Eva Farkas, Michael Kunzinger, and Roland Steinbauer,** On the foundations of nonlinear generalized functions I and II, 2001

728 **Laura Smithies,** Equivariant analytic localization of group representations, 2001

727 **Anthony D. Blaom,** A geometric setting for Hamiltonian perturbation theory, 2001

726 **Victor L. Shapiro,** Singular quasilinearity and higher eigenvalues, 2001

725 **Jean-Pierre Rosay and Edgar Lee Stout,** Strong boundary values, analytic functionals, and nonlinear Paley-Wiener theory, 2001

724 **Lisa Carbone,** Non-uniform lattices on uniform trees, 2001

723 **Deborah M. King and John B. Strantzen,** Maximum entropy of cycles of even period, 2001

722 **Hernán Cendra, Jerrold E. Marsden, and Tudor S. Ratiu,** Lagrangian reduction by stages, 2001

721 **Ingrid C. Bauer,** Surfaces with $K^2 = 7$ and $p_g = 4$, 2001

720 **Palle E. T. Jorgensen,** Ruelle operators: Functions which are harmonic with respect to a transfer operator, 2001

719 **Steve Hofmann and John L. Lewis,** The Dirichlet problem for parabolic operators with singular drift terms, 2001

718 **Bernhard Lani-Wayda,** Wandering solutions of delay equations with sine-like feedback, 2001

717 **Ron Brown,** Frobenius groups and classical maximal orders, 2001

716 **John H. Palmieri,** Stable homotopy over the Steenrod algebra, 2001

715 **W. N. Everitt and L. Markus,** Multi-interval linear ordinary boundary value problems and complex symplectic algebra, 2001

714 **Earl Berkson, Jean Bourgain, and Aleksander Pełczynski,** Canonical Sobolev projections of weak type $(1,1)$, 2001

713 **Dorina Mitrea, Marius Mitrea, and Michael Taylor,** Layer potentials, the Hodge Laplacian, and global boundary problems in nonsmooth Riemannian manifolds, 2001

712 **Raúl E. Curto and Woo Young Lee,** Joint hyponormality of Toeplitz pairs, 2001

711 **V. G. Kac, C. Martinez, and E. Zelmanov,** Graded simple Jordan superalgebras of growth one, 2001

710 **Brian Marcus and Selim Tuncel,** Resolving Markov chains onto Bernoulli shifts via positive polynomials, 2001

709 **B. V. Rajarama Bhat,** Cocylces of CCR flows, 2001

708 **William M. Kantor and Ákos Seress,** Black box classical groups, 2001

707 **Henning Krause,** The spectrum of a module category, 2001

706 **Jonathan Brundan, Richard Dipper, and Alexander Kleshchev,** Quantum Linear groups and representations of $GL_n(\mathbb{F}_q)$, 2001

705 **I. Moerdijk and J. J. C. Vermeulen,** Proper maps of toposes, 2000

704 **Jeff Hooper, Victor Snaith, and Min van Tran,** The second Chinburg conjecture for quaternion fields, 2000

(Continued in the back of this publication)

Tilings of the Plane, Hyperbolic Groups and Small Cancellation Conditions

MEMOIRS
of the
American Mathematical Society

Number 733

Tilings of the Plane, Hyperbolic Groups and Small Cancellation Conditions

Milé Krajčevski

ovember 2001 • Volume 154 • Number 733 (fourth of 5 numbers) • ISSN 0065-9266

American Mathematical Society
Providence, Rhode Island

2000 *Mathematics Subject Classification.*
Primary 20F65, 20F67; Secondary 05B45, 52C20, 57M05.

Library of Congress Cataloging-in-Publication Data

Krajcevski, Milé, 1954–
Tilings of the plane, hyperbolic groups, and small cancellation conditions / Milé Krajcevski.
p. cm. — (Memoirs of the American Mathematical Society, ISSN 0065-9266 ; no. 733)
"Volume 154, number 733 (fourth of 5 numbers)."
ISBN 0-8218-2762-6 (alk. paper)
1. Hyperbolic groups. 2. Tiling (Mathematics) 3. Cancellation theory (Group theory)
I. Title. II. Series.

QA3 .A57 no. 733
[QA174.3]
510 s—dc21
[512′.2] 2001034318

Memoirs of the American Mathematical Society

This journal is devoted entirely to research in pure and applied mathematics.

Subscription information. The 2001 subscription begins with volume 149 and consists of six mailings, each containing one or more numbers. Subscription prices for 2001 are $494 list, $395 institutional member. A late charge of 10% of the subscription price will be imposed on orders received from nonmembers after January 1 of the subscription year. Subscribers outside the United States and India must pay a postage surcharge of $31; subscribers in India must pay a postage surcharge of $43. Expedited delivery to destinations in North America $35; elsewhere $130. Each number may be ordered separately; *please specify number* when ordering an individual number. For prices and titles of recently released numbers, see the New Publications sections of the *Notices of the American Mathematical Society.*

Back number information. For back issues see the *AMS Catalog of Publications.*

Subscriptions and orders should be addressed to the American Mathematical Society, P. O. Box 845904, Boston, MA 02284-5904. *All orders must be accompanied by payment.* Other correspondence should be addressed to Box 6248, Providence, RI 02940-6248.

Memoirs of the American Mathematical Society is published bimonthly (each volume consisting usually of more than one number) by the American Mathematical Society at 201 Charles Street, Providence, RI 02904-2294. Periodicals postage paid at Providence, RI. Postmaster: Send address changes to Memoirs, American Mathematical Society, P. O. Box 6248, Providence, RI 02940-6248.

This publication is indexed in *Science Citation Index*®, *SciSearch*®, *Research Alert*®, *CompuMath Citation Index*®, *Current Contents*®/*Physical, Chemical & Earth Sciences.*
Printed in the United States of America.

∞ The paper used in this book is acid-free and falls within the guidelines established to ensure permanence and durability.
Visit the AMS home page at URL: `http://www.ams.org/`

10 9 8 7 6 5 4 3 2 1 06 05 04 03 02 01

Contents

Abstract

We consider a class of groups $\mathcal{T}$ which satisfies a generalized $C(4) - T(4)$ condition of small cancellation theory. To every group G in $\mathcal{T}$ we associate a finite set $T(G)$ of colored squares in the plane, Wang prototiles. We prove that the group $G \in \mathcal{T}$ is hyperbolic iff $T(G)$ is not a solvable set of prototiles, meaning one cannot tile the plane by copies of the prototiles in $T(G)$ respecting the color-matching conditions and allowing only translation of the prototiles in $T(G)$. The second result is that G has $\mathbb{Z} \times \mathbb{Z}$ as a subgroup iff $T(G)$ is a solvable periodic set of prototiles.

Received by the editor February 5, 1995.

Preface

After Olshanskii's paper [12], where he proves the claim of Gromov in [6] that almost every finitely presentable group is (word) hyperbolic and after the work of [3] on automatic groups, it was natural to further investigate and characterize these two classes of groups. In this paper we are interested in a class of finitely presentable groups $\mathcal{T}$ which satisfy a generalized C(4)-T(4) condition of the small cancellation theory by allowing certain order 2 generators. This class contains the class of finitely generated free groups and it is closed under free products. The significance of this class is underlined by its connection with a class of tiling problems introduced by Hao Wang in the 60's in connection with some formulas from propositional calculus.

The class $\mathcal{T}$ was discovered by the late Craig Squier, who raised the question of whether for this class of groups the hyperbolicity question is decidable. This means, given a group $G \in \mathcal{T}$ by its presentation P_G, is there an algorithm which has as input the presentation P_G and as output the answer to the question: "Is G hyperbolic or not?". His hope was that a group G in $\mathcal{T}$ is hyperbolic iff G does not contain a subgroup isomorphic to $\mathbb{Z} \times \mathbb{Z}$.

Although the whole work was initiated as an effort to produce such an algorithm, this question remains unanswered. It will be interesting to see if the example of J. Kari [8] of a perfect aperiodic set of prototiles provides the "missing link." The existence of such a set indicates that the hyperbolicity question even for this class of groups is most probably undecidable.

The work is organized as follows. The first chapter is introductory, reviewing some of the standard small cancellation conditions, adding a new notion of a $C(p,q)$ group (Definition 1.1.4). We think that by imposing a restriction of this type on the defining relators, one can further explore techniques of van Kampen diagrams. One can notice that the definition of the class $\mathcal{T}$ is such that in a (singular) disk diagram no two 4-cells have more than one edge in common (or the pieces in the presentation P_G of $G \in \mathcal{T}$ are of length at most one). Condition (C2) will assure that no interior vertex in a (reduced singular) disk diagram is a common vertex to two 4-cells and one 2-cell (i.e., has degree 3), or a common vertex of two 4-cells and two 2-cells. So, if an interior vertex v in a singular disk diagram has degree four, then v is incident to four 4-cells. One can notice that the existence of the defining relators of length two is not crucial to the connection between the class $\mathcal{T}$ and the class of perfect prototiles $\mathcal{P}$. Nevertheless, we included them for the sake of generality. Some authors adopt the convention that if a generator in a group G has order 2, the edges labeled by that generator in the Cayley graph $\Gamma(G,S)$ of G are not oriented. We do not follow this convention. The chapter ends with the most popular definition of a word hyperbolic group and with the Gromov's necessary condition for a group to be word hyperbolic.

Chapter 2 begins with a definition of C-geodesic. The idea for a C-geodesic comes from the fact that every group in the class $\mathcal{T}$ admits a finite, Noetherian and confluent string rewriting system. The rewriting rules are given by the following. For every relator (relation) of form $b^{\epsilon_b}a^{\epsilon_a}\beta^{-\epsilon_\beta}\alpha^{-\epsilon_\alpha}$ $(b^{\epsilon_b}a^{\epsilon_a} = \alpha^{\epsilon_\alpha}\beta^{\epsilon_\beta})$ we have a rewriting rule $b^{\epsilon_b}a^{\epsilon_a} \to \alpha^{\epsilon_\alpha}\beta^{\epsilon_\beta}$. For every defining relator x^2, $x \in A \cup B$, we have a rewriting rule $x^2 \to 1$. Together with the obvious rules $xx^{-1} \to 1$ and $x^{-1}x \to 1$ this will define a finite Noetherian and (locally) confluent rewriting system. Of course, we show that C-geodesics are geodesics and that every element $g \in G$ has a unique C-geodesic representative. In contrast with the work of [4] we take advantage of the explicit presentation of the groups in $\mathcal{T}$. An interesting feature of (singular) disk diagrams over a presentation P_G of $G \in \mathcal{T}$ is that whenever a singular disk diagram includes two geodesics on its boundary this diagram is uniquely determined (Theorems 2.2.2 and 2.2.3). Actually, there is an algorithm which has as input two geodesic words W, U representing the same element $g \in G$ and as output a singular disk diagram (M, ϕ, p, v_0) with $\phi(p) \equiv UW^{-1}$. Finally, in (2.2.1) we give a classification of simple C-geodesic triangles in a Cayley graph $\Gamma(G, S)$. Taking into account the shapes of geodesic digons (Corollary 2.2.6) we see that the size of a geodesic triangle is determined by the size of the so-called r-components of that triangle (2.2.1).

The main result in Chapter 2 is Theorem 2.2.4, which is used later (Theorem 8.5) to see a connection between periodic sets of prototiles T and the existence of $\mathbb{Z} \times \mathbb{Z}$ as a subgroup in the group $\mathcal{G}(T)$ associated with T. In order to simplify the proof of Theorem 2.2.4 we introduce the so-called component graph of a singular disk diagram (Def. 2.2.2) which, in some sense, captures the homotopy type of a given singular disk diagram.

Chapter 3 begins with a review of the tiling problem, mostly following [7] and [14]. By interpreting the relators in P_G for $G \in \mathcal{T}$ as prototiles, with every presentation P_G we associate a set of prototiles $\mathbb{T}(P_G)$ (3.3.1) and with every set of prototiles T we associate a group $\mathcal{G}(T)$ or more precisely a presentation $P_{\mathcal{G}(T)}$. One may notice that the definition of $\mathcal{G}(T)$ is pretty much straightforward, so it is to be expected that, unless we do not impose some additional restrictions on T, the problem of characterizing $\mathcal{G}(T)$ is hopeless. Those restrictions come in the form of "perfectness" (3.1.2) (which in some sense is suggested by the condition (C1) in the definition of $\mathcal{T}$) and "unique extendibility" (3.1.2). The latter condition significantly reduces the value of Theorem 3.1.2. There are relatively simple examples of non-solvable perfect sets of prototiles (which are not uniquely extendible) for which the corresponding group contains $\mathbb{Z} \times \mathbb{Z}$ as a subgroup (Example 3.1.4).

Finally, in Section 3.2 we give a characterization of the class of groups $\mathcal{T}$ by the class of perfect sets of prototiles (Prop. 8.3) and we answer (reformulate!) the question of hyperbolicity: a group G in $\mathcal{T}$ is hyperbolic iff the set of prototiles $\mathbb{T}(P_G)$ associated with the presentation P_G of G is not solvable (Theorem 3.2.1). Further, $\mathbb{Z} \times \mathbb{Z}$ is a subgroup of G iff $\mathbb{T}(G)$ is periodic (Theorem 3.2.2). The last two results are actually the main results of the whole work.

Acknowledgments

I would like to express my gratitude to Lee Mosher, who provided detailed comments and corrections of the entire manuscript which significantly improved the presentation of the material. Special thanks to Ross Geoghegan for his constant support and interest in this work which became part of my dissertation done at Binghamton University. I am thankful to Nataša Jonoska for enumerable suggestions and improvements of my mathematical style. Finally, I am thankful to the anonymous referee for pointing out numerous misprints and inaccuracies in an early version of this text.

CHAPTER 1

Introduction

1.1. Definitions

Given a set of symbols S we can associate the set S^{-1}, so called set of formal inverses of elements in S, consisting of all symbols of the form s^{-1}, $s \in S$. We say that the element s^{-1} is the *formal inverse* of the element $s \in S$. We will assume that $(s^{-1})^{-1} = s$, that is, S is the set of formal inverses of S^{-1}. The set $S \cup S^{-1}$ $(= S^{\pm})$ is called a *group alphabet* and its elements are called *letters.* A *word* over (in) the alphabet $S^{\pm}$ is a finite sequence $x_1, x_2, \ldots, x_n$ (or simply $x_1 x_2 \cdots x_n$) of letters, with $x_i \in S^{\pm}$. The number n is called the *length* of the word $X = x_1 x_2 \cdots x_n$ and is denoted by $|X|$. The empty word, denoted by 1, is the only word with length 0. We say that a word $X = x_1 x_2 \cdots x_n$ is *freely reduced* if there is no subscript i such that the letters x_i and x_{i+1} are formal inverses of each other.

Let $F(S)$ denote the free group on S. By identifying S with the equivalence classes of one-letter words in $F(S)$ we can regard S as a generating subset of $F(S)$ and we will refer to it as a *basis* for $F(S)$. At this point it is useful to define a *visual equality* of two words X and Y over the same alphabet as letter-for-letter equality of words of same length (not just equality as elements of a free group). If X is visually equal to Y we write $X \equiv Y$. For example $abb^{-1}a^{-1} = 1$ in the free group $F(\{a, b\})$, but $abb^{-1}a^{-1} \not\equiv 1$. If W is a word over the alphabet $S^{\pm}$ and $W \equiv XW_1Y$ we say that X is a *prefix* of W, Y is a*suffix* of W and W_1 is a *subword* of W. If at least one of X and Y are non-empty we call W_1 *a proper subword* of W.

PROPOSITION 1.1.1. ([11, p. 26], [9, p. 1]) *For any group G with an arbitrary set of generators $\{g_i\}_{i \in I}$, there exists a surjective homomorphism φ from the free group $F(S)$ with basis $S = \{s_i\}_{i \in I}$ onto G such that $\varphi(s_i) = g_i$, for all $i \in I$.* □

When a generating set $\{g_i\}_{i \in I}$ of elements in G is chosen, the homomorphism φ in Proposition 1.1.1 is called a *presentation* of the group G. Obviously, each presentation depends of the choice of the generating set for G. By the First Isomorphism Theorem, we have that $G \cong F(S)/Ker\varphi$. So any group G can be represented as a quotient of a free group by some normal subgroup N. What is the most "effective" way to specify this normal subgroup? Recall that for a given subset $\mathcal{R}$ of G, the normal closure $\mathcal{R}^G$ of $\mathcal{R}$ in G is the smallest normal subgroup of G containing $\mathcal{R}$. Its elements can be expressed as products of finitely many conjugates of elements of $\mathcal{R}$, i.e.,

$$g \in \mathcal{R}^G \Leftrightarrow g = \prod_{i=1}^{n} g_i r_i^{\pm 1} g_i^{-1}, \quad r_i \in \mathcal{R}, \quad g_i \in G.$$

Let $\mathcal{R}$ be fixed subset of $F(S)$. We define elementary transformations on words W in $F(S)$ which will leave fixed each coset of $\mathcal{R}^F$ as:

(1) The cancellation of two neighboring mutually inverse letters in W.
(1′) The inverse transformation of (1).
(2) Replacement of a subword $X \equiv X_1 R^{\pm 1} X_2$ in W by $X_1 X_2$, where $R \in \mathcal{R}$ and X_1, X_2 are any two words over the group alphabet $S^{\pm}$.
(2′) The inverse transformation of (2).

Two words X and Y over $S^{\pm}$ are called *$\mathcal{R}$-equivalent* if we can obtain Y from X by applying a finite sequence of elementary transformations of the above type. This introduces an equivalence relation on $F(S)$ and we say that the relation $W = V$ is *$\mathcal{R}$-deducible*, or that it can be deduced from the set of relations $\{R = 1 \mid R \in \mathcal{R}\}$, when W and V are $\mathcal{R}$-equivalent.

Finally, we say that $W = 1$, $W \in F(S)$ is a *consequence* of a set of relations $\{R = 1 \mid R \in \mathcal{R}\}$ or a set of relators $\{R \mid R \in \mathcal{R}\}$ if for any group G with fixed generating set $\{g_i\}_{i \in I}$ (hence with fixed homomorphism $\varphi : F(S) \to G$ such that $\varphi(s_i) = g_i$) the value of W under φ is 1 in G, provided that the values of all $R \in \mathcal{R}$ are equal to 1 in G.

PROPOSITION 1.1.2. ([11, p. 29]) *The following are equivalent:*
(i) $W = 1$ *is a consequence of* $\{R = 1 \mid R \in \mathcal{R}\}$, $\mathcal{R} \subset F(S)$.
(ii) W *belongs to the normal closure* $N = \mathcal{R}^F$.
(iii) $W = 1$ *is* $\mathcal{R}$*-deducible.*

□

By a *relator* among the elements $\{g_i\}_{i \in I}$ of a group G, we always mean a word W in $F(S)$ (where $S = \{s_i\}_{i \in I}$) whose value equals 1 in G with respect to a given presentation $\varphi : F(S) \to G$ such that $\varphi(s_i) = g_i$. In writing down a relation we abuse the notation and make no distinction between elements of G and words over the group alphabet $S^{\pm}$. A set of relations $\{R = 1 \mid R \in \mathcal{R}\}$ or a set of relators $\mathcal{R}$, is said to define a group G generated by the set $\{g_i\}_{i \in I}$ if any relation among the generators g_i in G is a consequence of $\{R = 1 \mid R \in \mathcal{R}\}$. Any $R \in \mathcal{R}$ is called a *defining relator* and $R = 1$ is *a defining relation* for G. By Proposition 1.1.2 and by the First Isomorphism Theorem, $G \cong F(S)/\mathcal{R}^{F(S)}$, therefore, the defining relations do indeed define the group G up to isomorphism. We write

$$(*) \qquad G = \langle S|\mathcal{R}\rangle \text{ or } G = \langle S \mid R = 1, R \in \mathcal{R}\rangle$$

and call (*) together with the homomorphism $\varphi : F(S) \to G$ a *presentation* for G. If S and $\mathcal{R}$ are finite, we say that G is *finitely presentable* or admits a finite presentation $P_G = \langle S \mid \mathcal{R}\rangle$.

Let $P_G = \langle S \mid \mathcal{R}\rangle$ be a given presentation for the group G and U be a set of symbols disjoint from S. Let $\{W_u(S) \mid u \in U\}$ be a set of words in the group alphabet $S^{\pm}$. Define $S' = S \cup U$ and $\mathcal{R}' = \mathcal{R} \cup \{uW_u(S)^{-1} \mid u \in U\}$. It can be shown that $P'_G = \langle S'|\mathcal{R}'\rangle$ is a presentation for the same group G. The passage from the presentation P_G to the presentation P'_G is called adding new generators U. The inverse procedure is called deleting generators of U.

Let $\mathcal{Q}$ be a set of relators which are consequences of the defining relators $\mathcal{R}$, possibly including trivial relators. Define $S'' = S$ and $\mathcal{R}'' = \mathcal{R} \cup \mathcal{Q}$. Again, it can be shown that $P''_G = \langle S'' \mid \mathcal{R}''\rangle$ is a presentation for the same group G and the passage from P_G to P''_G is called adding consequences of defining relators. The inverse procedure is called deleting redundant relators.

The above two procedures are known as *Tietze transformations*. If U or $\mathcal{Q}$ are finite, we call the procedures *finite Tietze transformations*.

THEOREM 1.1.1. ([10, Th. 1.5], [9, Prop. 2.1]) (Tietze transformation theorem) *Two presentations define isomorphic groups iff one can be transformed into the other by a finite sequence of Tietze transformations. If both presentations are finite, then only finitely many Tietze transformations are needed.* □

1.1.1. Small Cancellation Conditions. We recall a few standard definitions from small cancellation theory (see [9], [13], [11]). Let G be a group given by a presentation $P_G = \langle S \mid \mathcal{R} \rangle$. We'll require that every relator $R \in \mathcal{R}$ be freely reduced and cyclically reduced (that is, all cyclic permutations of defining relators are freely reduced).

The *symmetrization* $\mathcal{R}_*$ of $\mathcal{R}$ consists of all distinct cyclic permutations of defining relators and their inverses. Using the Tietze Transformation Theorem it is easy to show that $\langle S \mid \mathcal{R} \rangle$ and $\langle S \mid \mathcal{R}_* \rangle$ are presentations of the same group. So whenever a presentation $P_G = \langle S \mid \mathcal{R} \rangle$ of a group G is given, we may assume that $\mathcal{R}$ is symmetrized, that is $\mathcal{R} = \mathcal{R}_*$.

If $R_1 \equiv XY_1$ and $R_2 \equiv XY_2$ are two distinct words in $\mathcal{R}_*$ which have the same prefix X, X is called a *piece* (relative to $\mathcal{R}$).

DEFINITION 1.1.1. Let λ be a real number $0 < \lambda \leq 1$. We say that G is a $C'(\lambda)$ group if it admits a presentation $P_G = \langle S \mid \mathcal{R} \rangle$ in which for every $R \in \mathcal{R}$ such that $R \equiv XY$ with X a piece, we have that $|X| < \lambda |R|$.

Having that $\mathcal{R}$ is symmetrized, the condition $C'(\lambda)$ refers not only to common prefixes of words in $\mathcal{R}$ but also to common suffixes of words in $\mathcal{R}$. Therefore, the previous definition can be reformulated as:

> If $R_1 \equiv Z_1 X^{-1}$ and $R_2 \equiv XZ_2$ are two elements of $\mathcal{R}$ not inverse to each other, with X a piece, then $|X| < \lambda \min(|R_1|, |R_2|)$.

That is, in forming the product $R_1 R_2$, only a "small" part of each factor of that product (determined by λ) is canceled. This explains the use of the term "small cancellation."

EXAMPLE 1.1.1. Consider a presentation:

$$\langle a_1, b_1, \ldots, a_n, b_n \mid a_1 b_1 a_1^{-1} b_1^{-1} \cdots a_n b_n a_n^{-1} b_n^{-1} \rangle$$

of the fundamental group of a closed orientable surface M_n of genus n. The corresponding symmetrized presentation contains $8n$ relators: $4n$ cyclic conjugates of the given relator and their inverses. It is easy to see that the empty word and all single letter words are pieces, but no word of length greater than or equal to 2 is a piece. So, it follows that $\pi_1(M_n)$ satisfies the condition $C'(\lambda)$ for any $\lambda > 1/4n$.

DEFINITION 1.1.2. Let p be a natural number, $p \geq 2$. We say that G is a $C(p)$ group if G admits a presentation $P_G = \langle S \mid \mathcal{R} \rangle$ in which no element of $\mathcal{R}$ is a product of fewer than p pieces.

EXAMPLE 1.1.2.

1. If $G = \mathbb{Z} \oplus \mathbb{Z}$ and $P_G = \langle a, b \mid aba^{-1}b^{-1} \rangle$ then G is a $C(4)$ group since the empty word and the single letter words a, a^{-1}, b, b^{-1} are all possible pieces relative to $\mathcal{R} = \{aba^{-1}b^{-1},\ ba^{-1}b^{-1}a,\ a^{-1}b^{-1}ab,\ b^{-1}aba^{-1},\ bab^{-1}a^{-1},\ ab^{-1}a^{-1}b,\ b^{-1}a^{-1}ba,\ a^{-1}bab^{-1}\}$.
2. The free group on n generators, F_n, given with the standard presentation $\langle a_1, \ldots, a_n \mid \rangle$, is a $C(p)$ group for all $p \geq 2$.

NOTE 1.1.1. It is clear that the condition $C'(\lambda)$ implies the condition $C([\frac{1}{\lambda}]+1)$. Following [11, p. 133], A.I. Gol'berg has shown that even for $\lambda > \frac{1}{5}$ the condition $C'(\lambda)$ (and also $C(5)$) is essentially insignificant, i.e., any group satisfies this condition. That is why conditions $C'(\lambda)$ and $C(p)$ are used in combination with the so-called $T(q)$ condition.

DEFINITION 1.1.3. Let q be a natural number $q \geq 3$. We say that G is a $T(q)$ group if G admits a presentation $P_G = \langle S \mid \mathcal{R} \rangle$ which satisfies the following so called $T(q)$ condition:

For every $l \in \{3, 4, \ldots, q-1\}$ and every sequence $(R_1, R_2, \ldots, R_l)$ of elements of $\mathcal{R}$, the following implication holds:
If $R_1 \not\equiv R_2^{-1}, \ldots, R_{l-1} \not\equiv R_l^{-1}$ and $R_l \not\equiv R_1^{-1}$ then at least one of the products $R_1R_2, \ldots, R_{l-1}R_l$, R_lR_1 is freely reduced.

Note that, by definition, every group is a $T(3)$ group. Slightly less trivial is the example of $\mathbb{Z} \oplus \mathbb{Z}$ with a presentation $\langle a, b \mid bab^{-1}a^{-1} \rangle$ which satisfies the $T(4)$ condition i.e., $\mathbb{Z} \oplus \mathbb{Z}$ is a $T(4)$ group.

In addition to the previous classical definitions in small cancellation theory we will need a new one:

DEFINITION 1.1.4. (compare [4]) Let p and q be natural numbers $2 \leq p < q$. We say that G is a $C(p,q)$ group if it admits a presentation $P_G = \langle S \mid \mathcal{R} \rangle$ such that for every element $R \in \mathcal{R}$ which is a product of exactly n pieces, $n = p$ or $n = q$.

Obviously, every $C(p,q)$ group is also a $C(p)$ group.

EXAMPLE 1.1.3.

1. The group of braids with n strings, B_n given with its standard presentation $\langle \sigma_1, \ldots, \sigma_{n-1} \mid \sigma_j\sigma_{j+1}\sigma_j\sigma_{j+1}^{-1}\sigma_j^{-1}\sigma_{j+1}^{-1} (1 \leq j \leq n-2), [\sigma_j, \sigma_k] (1 \leq j < k-1 \leq n-2) \rangle$ is a $C(4,6)$ group.
2. Direct product of two cyclic groups of order 3, $\mathbb{Z}_3 \times \mathbb{Z}_3$, given with the presentation $\langle c, d \mid cdc^{-1}d^{-1}, c^3, d^3 \rangle$ is a $C(3,4)$ group.

1.2. The class $\mathcal{T}$

In what follows we will be interested in a class of groups $\mathcal{T}$ such that every group $G \in \mathcal{T}$ admits the following finite presentation P_G:

generators: they are partitioned in two non-empty disjoint sets A and B

relators: there are two kinds of defining relators

(i) Defining 4-relators, which constitute the set $\mathcal{R}_4$, consisting of words in the group alphabet $A^{\pm} \cup B^{\pm}$ of the form $b^{\epsilon_b}a^{\epsilon_a}\beta^{\epsilon_\beta}\alpha^{\epsilon_\alpha}$ where $a, \alpha \in A$; $b, \beta \in B$; and $\epsilon_a, \epsilon_b, \epsilon_\alpha, \epsilon_\beta = \pm 1$.

(ii) Defining 2-relators, which constitute the set $\mathcal{R}_2$, consisting of words of the form x^2 where x is a symbol in A or B.

which satisfy the following condition (C):

(C1) No two elements in $\mathcal{R}_{4*}$ have the same prefix of length 2.

(C2) If $\xi_1^{\epsilon_1}\xi_2^{\epsilon_2}\xi_3^{\epsilon_3}\xi_4^{\epsilon_4}$ and $\eta_1^{\nu_1}\eta_2^{\nu_2}\eta_3^{\nu_3}\eta_4^{\nu_4}$ are two words in $\mathcal{R}_{4*}$ such that $\xi_1 \equiv \eta_1$ then $\xi_2^{-\epsilon_2}\eta_2^{\nu_2} \notin \mathcal{R}_{2*}$.

EXAMPLE 1.2.1.

1. Finitely generated free groups or direct products of finitely generated free groups are among the trivial examples of groups in $\mathcal{T}$.

2. Let G be given by the presentation $P_G = \langle a_1, \alpha, b, \beta \mid ba_1\beta^{-1}\alpha^{-1}, \beta a_1 b^{-1}\alpha, a_1^2\rangle$. By applying some elementary Tietze transformations (deleting a_1 from the generators and adding a new generator $a = b^{-1}\alpha\beta$) we will get a different presentation P'_G of the same group, $P'_G = \langle a, \alpha, \beta \mid a^2\rangle$. Obviously $G \cong \mathbb{Z}_2 * \mathbb{Z} * \mathbb{Z}$ is in $\mathcal{T}$, but the presentation P_G does not satisfy the condition (C2), namely $ba_1\beta^{-1}\alpha^{-1}$ and $ba_1^{-1}\beta^{-1}\alpha^{-1}$ are both in $\mathcal{R}_{4*}$ and $a_1^{-2} \in \mathcal{R}_{2*}$. This should illustrate some of the difficulties in giving another ("presentation free" or "algebraic") characterization of the class $\mathcal{T}$.
3. The infinite dihedral group D_∞ can be defined as the group of 2×2 matrices:
$$\begin{pmatrix} \epsilon & k \\ 0 & 1 \end{pmatrix} \quad \text{where } \epsilon = \pm 1 \text{ and } k \in \mathbb{Z}$$
under matrix multiplication. Then D_∞ has a presentation $P_{D_\infty} = \langle a, b \mid baba, b^2\rangle$ under the mapping:
$$a \mapsto \begin{pmatrix} 1 & -1 \\ 0 & 1 \end{pmatrix}, \quad b \mapsto \begin{pmatrix} -1 & 0 \\ 0 & 1 \end{pmatrix}.$$
Defining 4-relator $baba$ and 2-relator b^2 satisfy the condition (C). Therefore $D_\infty \in \mathcal{T}$.
4. We will start with the following presentation $P_G = \langle S' | \mathcal{R}' \rangle$ of a group G, where the generators S' and the set of relators $\mathcal{R}$ are given by

generators: $\begin{cases} a_{11}, & a_{12}, & a_{21}, & a_{22} \\ b_{11}, & b_{12}, & b_{21}, & b_{22} \\ \alpha_1, & \alpha_2, & \alpha_3, & \alpha_4 \\ \beta_1, & \beta_2, & \beta_3, & \beta_4 \end{cases}$

relators: $\begin{cases} \beta_1\alpha_1 b_{11}^{-1} a_{12}^{-1} \\ \beta_2\alpha_1 b_{12} a_{21} \\ \beta_2\alpha_2 b_{21}^{-1} a_{22}^{-1} \\ \beta_3 a_{12} b_{22}^{-1} \alpha_2^{-1} \\ \beta_3 a_{11}^{-1} b_{12}^{-1} \alpha_3 \\ \beta_4 a_{22} b_{11} \alpha_3 \\ \beta_4 a_{21}^{-1} b_{22} \alpha_4 \\ \beta_1 \alpha_4^{-1} b_{21}^{-1} a_{11} \end{cases}$

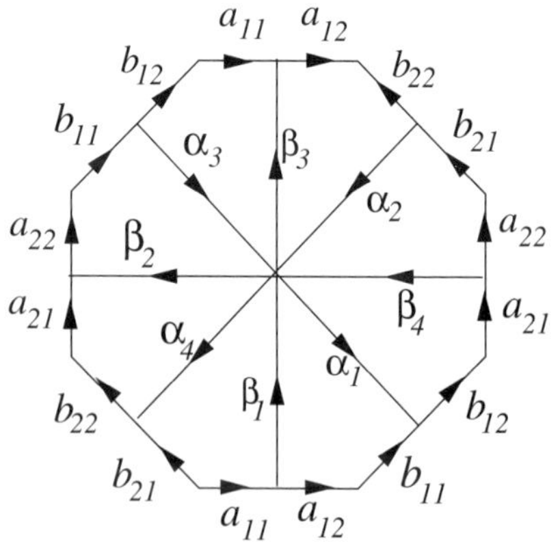

It is not dificult to see that G belongs to $\mathcal{T}$. Using Tietze transformations (deleting α_j, $j = 1, 2, 3, 4$ and $\beta_2, \beta_3\beta_4$, then adding new generators $a_i = a_{i1}a_{i2}$ and $b_i = b_{i1}b_{i2}$, $i = 1, 2$ and again deliting a_{i2}, b_{i2}) we get the following presentation $P_G = \langle S \mid \mathcal{R}\rangle$ of G:
$$\langle a_1, a_2, a_{11}, a_{21}, b_1, b_2, b_{11}, b_{21}, \beta_1 \mid a_1 b_1 a_2 b_2 a_1^{-1} b_1^{-1} a_2^{-1} b_2^{-1}\rangle.$$
Therefore G is a free product of $\pi_1(M_2)$, the foundamental group of two-dimensional orientable manifold M_2, with $\mathbb{Z} * \cdots * \mathbb{Z}$ $(2 \cdot 2 + 1$ copies of $\mathbb{Z})$. Of course, the previous example can be generalized and we can see that for any $g \geq 2$, the group $\pi_1(M_g) * \left(\underset{2g+1}{*} \mathbb{Z}\right)$ belongs to $\mathcal{T}$.
5. It follows easily from the definition of $\mathcal{T}$ that if $G_i \in \mathcal{T}$, $i = 1, 2$, then $G_1 * G_2$ is also in $\mathcal{T}$, that is the class $\mathcal{T}$ is closed under taking free products.

A justification for considering this class of groups will become clear later when we see a connection between $\mathcal{T}$ and "the class of perfect tiles".

It is clear that every $G \in \mathcal{T}$ is a $C(2,4)$ group, but it is not so obvious that every $G \in \mathcal{T}$ is $T(4)$ group.

PROPOSITION 1.2.1. *Every $G \in \mathcal{T}$ is a $T(4)$ group.*

PROOF. Let $P_G = \langle A, B \mid \mathcal{R}_4, \mathcal{R}_2 \rangle$ be a presentation of $G \in \mathcal{T}$ and let R_i, $i = 1,2,3$ be three elements in $\mathcal{R}_* = \mathcal{R}_{4*} \cup \mathcal{R}_{2*}$ such that $R_1 \not\equiv R_2^{-1}$, $R_2 \not\equiv R_3^{-1}$, $R_3 \not\equiv R_1^{-1}$. Assume that none of the products R_1R_2, R_2R_3, R_3R_1 is freely reduced. Note that it is not possible all of R_i to be in $\mathcal{R}_{4*}$ because for every element in $\mathcal{R}_{4*}$, the beginning letter and ending letter are from different sets of symbols. So, the only possibility is that two of R_i's are in $\mathcal{R}_{4*}$ and one of them in $\mathcal{R}_{2*}$. Assume that $R_1 \equiv b^\epsilon a_1^{\epsilon_1} \beta_1^{\nu_1} \alpha^\nu$, $R_2 \equiv \alpha^{-\nu}\alpha^{-\nu}$, $R_3 \equiv \alpha^\nu b_3^{\epsilon_3} a_3^{\nu_3} b^{-\epsilon}$. In this case $R_1^* = (a_1^{\epsilon_1} \beta_1^{\nu_1} \alpha^\nu b^\epsilon)^{-1}$ and $R_3^* = b^{-\epsilon} \alpha^\nu b_3^{\epsilon_3} a_3^{\nu_3}$ are two elements in $\mathcal{R}_{4*}$ which do not satisfy the condition (C2). This contradicts the assumption that $G \in \mathcal{T}$. So every $G \in \mathcal{T}$ is a $T(4)$ group. □

1.3. Singular Disk Diagrams, Reduced Diagrams and Cayley Graphs

It was van Kampen who first introduced the notion of a diagram in group theory. Later (in the mid-sixties) Roger Lyndon used the same ideas to initiate a geometric study of small cancellation conditions, where consequences of the cancellation hypothesis on G are studied by the properties of so-called "disk diagrams" over a given presentation P_G of G. We will illustrate the idea of a diagram with a simple example.

EXAMPLE 1.3.1. If relations $a^2 = 1$ and $ba = ab$ hold in some group G it obviously follows that $a^{-1}ba^3b^{-1} = 1$ also holds in G. This deduction is reflected in Figure 1.1.

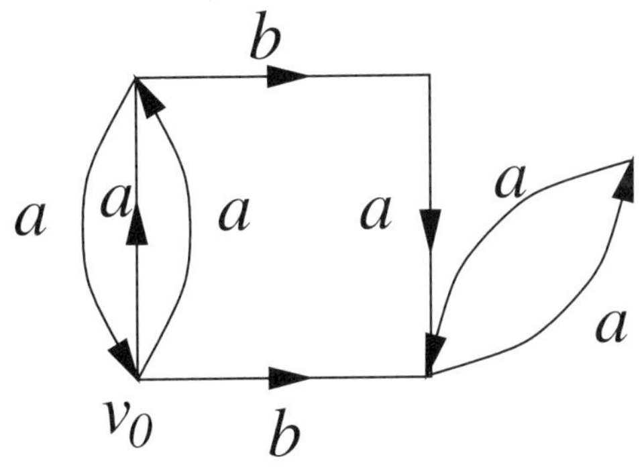

FIGURE 1.1

The Figure 1.1 represents a planar two-dimensional complex M with oriented edges labeled a or b, such that the label of every "simple boundary path" (see definition below) of a cell of M is an element of $\{a^2, bab^{-1}a^{-1}\}_*$. The boundary path p, beginning and ending at v_0 and taken clockwise, will represent the relator $a^{-1}ba^3b^{-1}$ which is a consequence of the relators a^2 and $bab^{-1}a^{-1}$. How was the complex M obtained?

By Proposition 1.1.2 an equality of the form

$$a^{-1}ba^3b^{-1} = \prod_{i=1}^{n} U_i R_i^{\pm 1} U_i^{-1} \tag{1.3.1}$$

must hold in the free group $F = F(a, b, \dots)$, where $U_i \in F$ and R_i is a defining relator (that is a^2 or $bab^{-1}a^{-1}$ in our example). Indeed, in this case we may have

$$a^{-1}ba^3b^{-1} = (1 \cdot a^{-2} \cdot 1)(1 \cdot a^2 \cdot 1)(a^{-1} \cdot bab^{-1}a^{-1} \cdot a)(b \cdot a^2 \cdot b^{-1}). \tag{1.3.2}$$

We represent the factors in this product (right hand side of (1.3.2)) as a wedge of four "tailed disks" (with possibly empty tails) (see Figure 1.2).

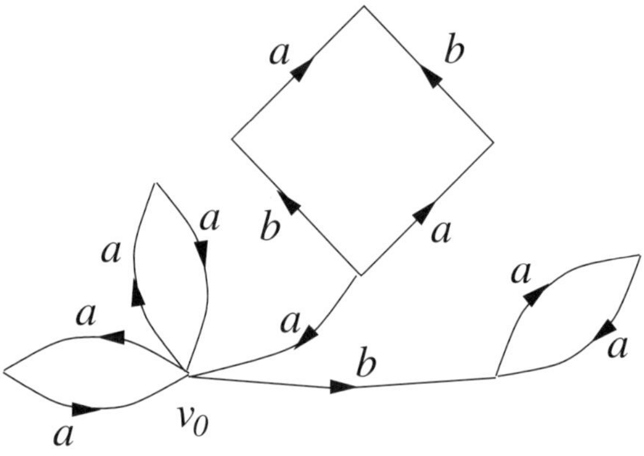

FIGURE 1.2

The circumferences of the disks are labeled such that we can read $(bab^{-1}a^{-1})^{\pm 1}$ or $a^{\pm 2}$ and the i-th tail, $i = 1, 2, 3, 4$ is labeled with the word U_i (the empty word, a or b). Then going (in clockwise direction) around, starting and ending with the vertex v_0, we read the right hand side of the equation (1.3.2). In order to obtain the left hand side of (1.3.2) it remains to carry out some cancellations, which are done by successively pasting together pairs of neighboring edges on the boundary with the same label and opposite orientation (see Figure 1.3).

Similarly we can construct a diagram for any relator, that is for any consequence of the defining relators, although different diagrams for the same consequence $W = 1$ may not be cell equivalent.

Following [9] we can formalize and generalize the previous construction as follows.

1.3.1. Relators as labels on boundaries of maps. By a *vertex* we mean a point in the Euclidean plane $\mathbb{E}$. An *edge* is a bounded subset of $\mathbb{E}$ homeomorphic to the open unit interval. A *cell* or a *face* is a bounded subset of $\mathbb{E}$ homeomorphic with the open unit disk. A *map* is a finite collection, M, of vertices, edges and cells which are pairwise disjoint and satisfy the following:

(i) If e is an edge of M there are vertices v and u (not necessarily distinct) in M such that $\bar{e} =$ the closure of $e = e \cup \{v\} \cup \{u\}$.
(ii) The boundary $\dot{B}$ of a cell B in M is connected and there is a set of edges $e_1, e_2, \dots, e_n$ in M such that $\dot{B} = \bar{e}_1 \cup \dots \cup \bar{e}_n$.

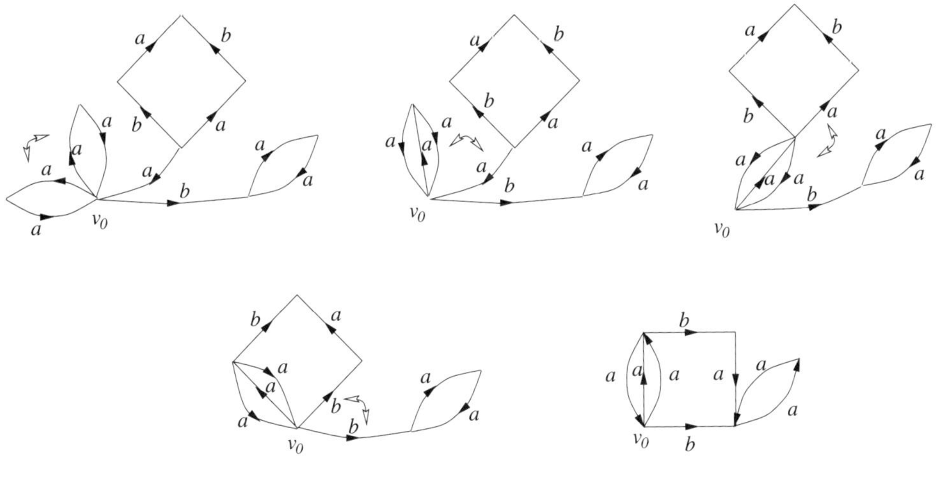

FIGURE 1.3

(iii) For any two distinct vertices u and v there exist vertices $u_0 = u, u_1, \ldots, u_{k+1} = v$ $(k \geq 0)$ and edges $e_1, \ldots, e_{k+1}$ such that $u_i = \bar{e}_i \cap \bar{e}_{i+1}$ $(1 \leq i \leq k)$.

Vertices v and u in (i) are called endpoints of the edge e. A *closed edge* is an edge together with its endpoints. We shall consider the edges and cells of M to be oriented. Thus, if e is an oriented edge, running from the endpoint v_1 to the endpoint v_2, the vertex v_1 is called the *initial* vertex and the vertex v_2 the *terminal* vertex of e. The oppositely oriented edge or inverse of e, denoted by e^{-1}, runs from v_2 to v_1 and consists of the same set of points as e. A *path* p is a finite sequence of oriented closed edges $e_1, e_2, \ldots, e_n$ (or simply $e_1 e_2 \cdots e_n$) such that the initial vertex of e_{i+1} is the terminal vertex of e_i, $i = 1, \ldots, n-1$. The endpoints of p are the initial vertex of e_1 and the terminal vertex of e_n. We say that n is the length of p $(n \geq 0)$. A *closed path* or a *cycle* is a path in which the endpoints coincide. A *reduced path* is a path which doesn't contain a successive pair of edges of the form ee^{-1}. A reduced path $e_1 \cdots e_n$ is *simple* if for $i \neq j$, e_i and e_j have different initial vertices. We can orient the cells of M and the components of M^c (= complement of M in $\mathbb{E}$) such that in traversing the boundaries of the cells of M and the boundary components of M^c each edge is traversed twice, once in each of its possible orientations.

Let B be an oriented cell of M and let the boundary edges of B be oriented in accordance with the orientation of B. By a *boundary cycle* of B we mean any cycle of minimal length which includes all the edges on the boundary $\dot{B}$ of B. And finally, when M is simply connected, we define a *boundary cycle* of M to be a cycle ω of minimal length which contains all edges of the boundary of M and doesn't cross itself, in the following sense:

> If e_i and e_{i+1} are consecutive edges of ω with e_i ending at vertex v, then e_i^{-1} and e_{i+1} are adjacent in the cyclically ordered set of all edges of M beginning at the vertex v (see Figure 1.4).

Going back to the previous example, we see how we can proceed in a general situation. Given a presentation $P_G = \langle S \mid \mathcal{R} \rangle$ of a group G and a freely reduced relator W_0 of $\langle S \mid \mathcal{R} \rangle$, by Proposition 1.1.2, W_0 is freely equivalent to W' which is

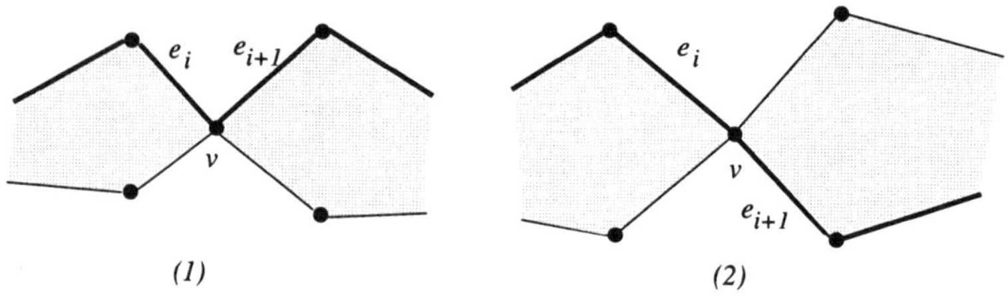

FIGURE 1.4. Figure (a) represents a part of a boundary cycle and the path indicated on Figure (b) is not part of any boundary cycle.

a product of conjugates of words in $\mathcal{R}_*$,

$$W' \equiv \prod_{i=1}^{n} U_i R_i^{\pm 1} U_i^{-1}.$$

We can assume that each factor $U_i R_i^{\pm 1} U_i^{-1}$ in this product is freely reduced but W' may not be freely reduced. We realize W' as the label of a boundary cycle p' of a pointed map M' which is a wedge of tailed disks.

We adopt the convention that a boundary cycle of a map M is always traversed clockwise. If W' as a label of a boundary cycle beginning at the vertex v_0 is freely reduced, we have obtained the desired diagram. If not, we can reduce the label W' of p' by sewing sub-paths $e_i e_{i+1}$ of p' of two consecutive edges whose labels are inverse to each other.

1.3.2. Singular Disk Diagrams. To summarize (see [13, pg. 232–233] or [9], for a discussion of this process) for a given presentation $P_G = \langle S \mid \mathcal{R} \rangle$ of a group G and a given freely reduced relator W_0 we can construct an oriented pointed map M, a boundary cycle p of M and a labeling function ϕ with the following properties:

(i) The space underlying M is homeomorphic to a simply connected closed subset of the plane.
(ii) ϕ associates to every oriented edge e of M a letter from $S^{\pm}$. Moreover $\phi(e^{-1}) = [\phi(e)]^{-1}$ for all oriented edges e of M.
(iii) The label of every boundary cycle of a cell of M is an element of $\mathcal{R}_*$.
(iv) The boundary cycle p starting and ending at the base point v_0 taken in the clockwise direction has a label of the given relator W_0.

The quadruple (M, ϕ, p, v_0) or simply M, when ϕ, p and v_0 are understood, is called *a singular disk diagram over* P_G with a boundary cycle p.

1.3.3. Reduced Disk Diagrams. The set of vertices, edges and cells (faces) of M will be denoted by $M^{(0)}$, $M^{(1)}$ and $M^{(2)}$ respectively. A vertex (edge) $v \in M^{(0)}$ ($e \in M^{(1)}$) is called *exterior* if it lies (is contained) on the boundary $\dot{M}$ of M; otherwise it is called *interior*. A cell $B \in M^{(2)}$ is *exterior* if one of the edges of $\dot{B}$ is exterior, otherwise is *interior*. A vertex $v \in M^{(0)}$ is a *cut* vertex if its removal disconnects the diagram. If C_M is the set of all cut vertices of a singular disk diagram M, the closure of every connected component of $M \setminus C_M$ is homeomorphic to either a disk or a closed interval. In the former case we call it a *disk component* of M. A singular disk diagram with no cut vertices is called *a disk diagram.* A *subdiagram* M' of an arbitrary singular disk diagram (M, ϕ, p, v_0) over

$P_G = \langle S \mid \mathcal{R} \rangle$ will be any singular disk diagram $(M', \phi|_{M'}, p', v_0')$ over the same presentation $P_G = \langle S \mid \mathcal{R} \rangle$, where $M' \subseteq M$ and $\phi|_{M'}$ is the restriction of ϕ on M'.

DEFINITION 1.3.1. A singular disk diagram (M, ϕ, p_0, v_0) is *reduced* if whenever an edge e is common to two cells B and B', the boundary paths p and p' of B and B' respectively, starting with e as their initial edge do not have identical labels.

A singular disk diagram which is not reduced is called *unreduced.* A pair of cells (B, B') is called *cancelable* if B and B' share a common edge and the boundary paths starting at that edge have same labels. If M is unreduced we can always cut out a cancelable pair of cells, together with their common boundary, pasting their uncommon boundaries in such a way that the new diagram represents the same relator and has $|M^{(2)}| - 2$ cells. By repeated application of this process, we can obtain a singular disk diagram with no cancelable pair of cells and is therefore reduced.

THEOREM 1.3.1. ([9], [13], or [11]) (The existence of reduced singular disk diagrams for freely reduced relators) *Let $P_G = \langle S \mid \mathcal{R} \rangle$ be a symmetrized presentation of a group G and let W_0 be a freely reduced relator in P_G. Then there is a reduced singular disk diagram (M, ϕ, p, v_0) over $\langle S \mid \mathcal{R} \rangle$ with $\phi(p) \equiv W_0$.*

NOTE 1.3.1. As a consequence of this theorem, we will assume that all singular disk diagrams are reduced.

Let v be a vertex in M. We define the *degree of a vertex* v, $d(v)$, to be the number of edges incident to v (counting twice each loop at v). We also define the *degree of a cell* B, written $d(B)$, as the number of edges making up $\dot{B}$, counted with multiplicity, that is, we take into account the number of appearances of the same edge.

THEOREM 1.3.2. ([9, p. 242]; [13, p. 239] (Geometric interpretation of the small cancellation conditions) *Let (M, ϕ, p, v_0) be a reduced singular disk diagram over a symmetrized presentation $P_G = \langle S|\mathcal{R} \rangle$ of a group G.*

(i) *If G, given with a presentation P_G, is a $C(p)$-group then every interior cell of M has degree at least p.*
(ii) *If G, given with a presentation P_G, is a $C(p,q)$-group then every interior cell of M has degree p or q.*
(iii) *If G, given with a presentation P_G, is a $T(q)$-group for some $q \geq 4$, then every interior vertex of M has degree at least q or degree 2.*

PROOF. The proof of (ii) follows the same lines as the proof of (i) and (iii) given in [9, p. 242]. □

Obviously, if B is a cell in a singular disk diagram (M, ϕ, p, v_0) over P_G, $G \in \mathcal{T}$, $\dot{B}$ is made up of either 2 or 4 edges. Therefore we can refer to B as a 2-cell or a 4-cell and denote it as B_2 or B_4 respectively. If $v \in M^{(0)}$ we define the 4-degree of v, $d_4(v)$, to be the number of 4-cells in M, counted with multiplicity, having v as a vertex on their boundaries. Similarly, we define the 2-degree of v, $d_2(v)$, to be the number of 2-cells in M, counted with multiplicity, having v as a vertex on their boundaries.

1.3.4. Cayley Graphs. For a definition of the Cayley graph of a group G we have the following set-up. Let $\{g_i\}_{i\in I}$ be a set of generators for the group G such that $1 \notin \{g_i\}_{i\in I}$ and let $\varphi : F(S) \to G$ ($\varphi(s_i) = g_i, i \in I$) be a given presentation of G. We define the Cayley graph $\Gamma(G, S)$ of the group G with respect to the generating subset $\{g_i\}_{i\in I}$ as the directed graph, having G as its set of vertices $V = V(\Gamma(G, S))$ and $G \times S^{\pm} \times G \supseteq \{(v, s^{\epsilon}, v\varphi(s)^{\epsilon}) \mid v \in V, s \in S\}$ as its set of directed edges $E = E(\Gamma(G, S))$. We require that $e = (v, s, v\varphi(s))$ and its inverse $e^{-1} = (v\varphi(s), s^{-1}, v)$ define the same unoriented edge and view each unoriented edge as an isomorph of the unit interval.

EXAMPLE 1.3.2. If G is the infinite cyclic group $(\mathbb{Z}, +)$ with generating subset $\{1\}$ and presentation $\varphi : F(\{a\}) \to \mathbb{Z}$ given by $\varphi(a) = 1$, (part of) the Cayley graph $\Gamma(G, \{a\})$ of G looks like the one on Figure 1.5 (1).

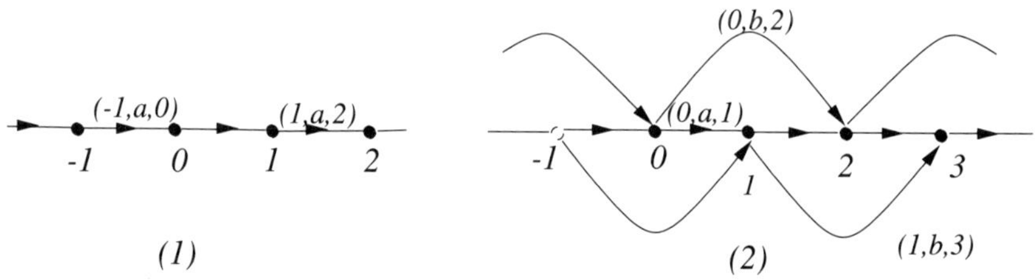

FIGURE 1.5

We see that the Cayley graph of G depends on the generating set for the group G. For example if $\mathbb{Z}$ was given by the presentation $\varphi : F(\{a, b\}) \to \mathbb{Z}$ with $\varphi(a) = 1$ and $\varphi(b) = 2$, i.e., $P_{\mathbb{Z}} = \langle a,\ b \mid b = a^2\rangle$ then (part of) the Cayley graph $\Gamma(\mathbb{Z}, \{a, b\})$ of $\mathbb{Z}$ is given on Figure 1.5 (2).

There is a natural action of G on the Cayley graph $\Gamma(G, S)$ by left multiplication $\mu : G \times \Gamma(G, S) \to \Gamma(G, S)$ given by

$$\mu(g, v) = gv, \text{ for all } g \in G \text{ and } v \in V(\Gamma(G, S))$$
$$\mu(g, (v, s^{\epsilon}, v\varphi(s)^{\epsilon})) = (gv, s^{\epsilon}, gv\varphi(s)^{\epsilon}), \text{ for all } g \in G, \text{ and } (v, s^{\epsilon}, v\varphi(s)^{\epsilon}) \in E$$

Sometimes we will refer to μ as *left translation.*

1.4. Geodesic Triangles in a Cayley Graph

DEFINITION 1.4.1. Let (X, d) be a metric space and let x and y be two points in X such that $d(x, y) = \lambda$. *A geodesic segment* in X with origin at x and ending at y is an isometry $\xi_{xy} : [0, \lambda] \to X$ such that $\xi_{xy}(0) = x$ and $\xi_{xy}(\lambda) = y$.

We will use the same notation for the image of such an isometry and call ξ_{xy} *a geodesic segment* with endpoints x and y. We are particularly interested in the case when X is the Cayley graph $\Gamma(G, S)$ of a finitely presented group G with a specified set of generators $\{g_i\}_{i\in I}$. In that case, $\Gamma(G, S)$ is equipped with the natural metric, given by taking each edge to have unit length. Then a geodesic for the element $g \in G$ is the shortest path ω in the Cayley graph from the vertex labeled 1 to the vertex labeled g. This is equivalent to taking the shortest word in the group alphabet $S^{\pm}$ representing the element g.

Given a word W over the group alphabet $S^{\pm}$, we can think of W as a mapping from the interval $[0, |W|]$ to $\Gamma(G, S)$ which maps the subintervals $[i-1, i]$, $i = 1, \ldots, |W|$, isometrically to edges of $\Gamma(G, S)$, such that $W(0) =$ identity element in G, and $W(|W|) = \varphi(W)$. In this case we will refer to W as a path $W(t)$ in the Cayley graph $\Gamma(G, S)$.

DEFINITION 1.4.2. (see [13]) If x, y and z are distinct points in (X, d) and if the geodesic segments ξ_{xy}, ξ_{yz} and ξ_{zx} have only their endpoints in common, the triangle Δxyz will be called a *simple geodesic triangle.*

If $\{x, y, z\}$ consists of two points, say $x \neq y = z$ and if geodesics ξ_{xy} and ξ_{yx} have only their endpoints in common, Δxyz will be called a *simple geodesic digon.*

Let ξ_{xy} and μ_{yz} be two geodesic segments in the Cayley graph $\Gamma(G, S)$ having $\{v_1, \ldots, v_n\}$ as a set of common vertices. We assume that $\xi_{xv_1} \subseteq \xi_{xv_2} \subseteq \cdots \subseteq \xi_{xy}$. If there is an i, $1 \leq i \leq n$, and j, $1 \leq j \leq n - i$, such that $\xi_{v_i v_{i+j}} = \mu_{v_{i+j} v_i}$ we say that the geodesic segments ξ_{xy} and μ_{yz} have a common geodesic segment.

We will say that the geodesic triangle Δxyz contains a simple geodesic triangle Δ' if there are vertices u, v and w on the geodesic segments ξ_{xy}, ξ_{yz} and ξ_{zx} respectively such that the geodesic triangle $\Delta' = \Delta' uvw$ with geodesic segments $\xi_{uv} \subseteq \xi_{xy}$, $\xi_{vw} \subseteq \xi_{yz}$ and $\xi_{wu} \subseteq \xi_{zx}$ is a simple geodesic triangle.

It is clear that every geodesic triangle in $\Gamma(G, S)$ is built up from finitely many geodesic segments, simple geodesic digons and at most one simple geodesic triangle (see Figure 1.6).

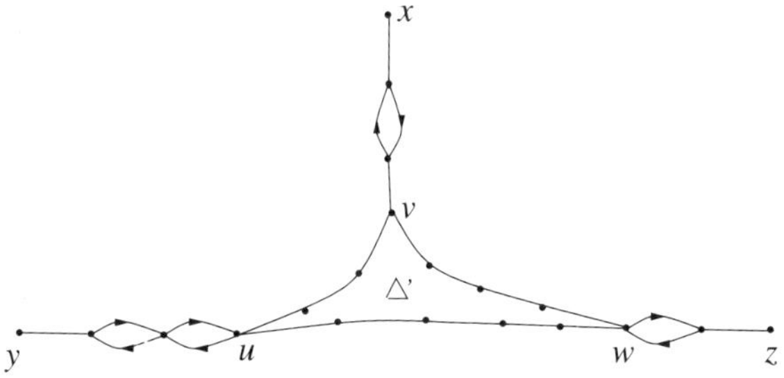

FIGURE 1.6. $\Delta' = \Delta' vuw$ is a simple geodesic triangle contained in the geodesic triangle $\Delta = \Delta xyz$.

1.4.1. The following construction will be applied to the geodesic triangles in $\Gamma(G, S) = X$ (see [13]).

Let p' be a path in X which starts and ends at 1. Let W_0 be the word in the group alphabet $S^{\pm}$ obtained by spelling out, in order, the second entries of the sequence of oriented edges $(g, s^{\epsilon}, v\varphi(s)^{\epsilon})$ making up p'. Assume that W_0 is cyclically reduced and let (M, ϕ, p, v_0) be a singular disk diagram over a presentation $P_G = \langle S | \mathcal{R} \rangle$ of G, with $\phi(p) = W_0$.

Then, there is a cellular map $\hat{f} : M^{(1)} \to \Gamma(G, S)$ with the following two properties:

(i) $\hat{f}$ maps the boundary path p onto p', in particular, sends the base point v_0 of M to $1 \in \Gamma(G, S)$.
(ii) $\hat{f}$ preserves the labels, i.e., an oriented edge e with label $\phi(e)^{\epsilon}$ is mapped to an oriented edge $(g, \phi(e), g\varphi(\phi(e))^{\epsilon})$.

This will imply that every path q in M, starting at v_0 maps onto a path in $\Gamma(G, S)$ starting at 1.

Note that two different edge paths in M starting at v_0 may be mapped into the same path in $\Gamma(G, S)$. This is so because in the Cayley graph, for each vertex $v \in V(\Gamma(G, S)$ and each symbol $s \in S$ there is one and only one edge $(v, s, s\varphi(s))$ starting at v and ending at $v\varphi(s)$, while in M there may be several edges e_i starting at v_0 and having the same label $\phi(e_i) = s$. On the other hand, a vertex v' in $M^{(0)}$ may be the initial vertex of several edges with the same label s.

NOTE 1.4.1. When (M, ϕ, p, v_0) is a disk diagram over P_G, $G \in \mathcal{T}$, representing a simple geodesic triangle Δxyz with geodesic sides $\xi_{xy}, \ldots$ we may define a metric d_M on M by taking $d_M(v_1, v_2)$ to be the length of a shortest path p_{12} with the initial vertex v_1 and the terminal vertex v_2. One can show that in this case the map $\hat{f}$ is an isometry therefore we can think of (M, ϕ, p, v_0) as a part of the Cayley graph $\Gamma(G, S)$ by identifying all interior vertices in M with the corresponding vertices in $\Gamma(G, S)$ and all interior edges in $M^{(1)}$ with the corresponding edges in $\Gamma(G, S)$.

DEFINITION 1.4.3. A simple geodesic digon or triangle Δxyz in $\Gamma(G, S)$ is *normalized* if the vertex x coincides with the vertex 1 in $\Gamma(G, S)$ and none of the vertices y and z coincides with 1.

Since G acts on the left by isometries, it is clear that by applying left translation, every triangle in $\Gamma(G, S)$ can be normalized.

A normalized simple geodesic digon or triangle Δxyz is uniquely determined by the triple of words (U_1, U_2, U_3) obtained by concatenation of the second entries of the oriented edges $(g, s^\epsilon, g\varphi(s)^\epsilon)$ which constitute the geodesic segments ξ_{xy}, ξ_{yz} and ξ_{zx} of Δ. If $\{x, y, z\}$ consists of only two points, say $x \neq y = z$, then Δxyz is a digon and U_2 is empty.

LEMMA 1.4.1 (see [13]). *If Δ is a normalized simple geodesic digon or triangle, the associated triple (U_1, U_2, U_3) of words has the following properties:*

(1) *Each U_i is freely reduced and the product $U_1U_2U_3$ is cyclically reduced.*
(2) *The product $U_1U_2U_3 = 1 \in G$ and if W is a nonempty subword of $U_1U_2U_3$ then $W \neq 1$.*
(3) *No word of length $< |U_i|$, $(i = 1, 2, 3)$ represents the same element in G as U_i does.* □

Let Δxyz be a geodesic triangle in $\Gamma(G, S)$ with geodesic sides ξ_{xy}, ξ_{yz} and ξ_{zx}. Let δ be a given non-negative constant. We say that Δxyz is a *δ-thin triangle* (or a δ-slim triangle, see [1]) if for any point v on ξ_{xy}, $\min\{d(v, \xi_{yz}), d(v, \xi_{zx})\} \leq \delta$. We say that $\Gamma(G, S)$ satisfies the *thin triangle condition* if there is a constant $\delta \geq 0$ such that all geodesic triangles in $\Gamma(G, S)$ are δ-thin.

DEFINITION 1.4.4. Let S be a finite set of generators for the group G. We say that G is (word) *hyperbolic* if the corresponding Cayley graph $\Gamma(G, S)$ of G satisfies the thin triangle condition.

It can be shown that this definition does not depend on the choice of the finite generating set S of G.

All finite groups, free groups of finite rank and all $C'(1/6)$ groups are examples of hyperbolic groups. Finally, we will be interested in the following necessary condition for hyperbolicity, due to Gromov.

THEOREM 1.4.1 (see [6],or [5]). *A hyperbolic group has no subgroup isomorphic to* $\mathbb{Z} \times \mathbb{Z}$. □

CHAPTER 2

Small Cancellation Theory of $\mathcal{T}$

2.1. C-geodesics and Chains

Let $P_G = \langle S \mid \mathcal{R}\rangle = \langle A, B \mid \mathcal{R}_4, \mathcal{R}_2\rangle$ be a presentation of a group $G \in \mathcal{T}$ and let $\Gamma(G, A \cup B)$ be the Cayley graph of G. Given two vertices v_1 and v_2 in the Cayley graph of G, there are finitely many geodesics in $\Gamma(G, A\cup B)$ having them as endpoints. Let $g_\Gamma(v_1, v_2)$ denote the set of all geodesics in $\Gamma(G, A\cup B)$ having v_1 as the initial vertex and v_2 as the terminal vertex. Our class $\mathcal{T}$ has the property that for every $G \in \mathcal{T}$ and any two vertices v_1 and v_2 in Γ the set $g_\Gamma(v_1, v_2)$ can be linearly ordered. Let $A = \{a_1, \ldots, a_k\}$ and $B = \{b_1, \ldots, b_l\}$. The elements of $A^\pm \cup B^\pm$ can be linearly ordered by

$$\begin{aligned} a_i^\epsilon < a_j^\nu \quad & \text{if } i<j \text{ or } i=j \text{ and } \epsilon = 1,\ \nu=-1 \\ a_i^\epsilon < b_j^\nu \quad & \text{for all } i,\ j,\ \epsilon \text{ and } \nu \\ b_i^\kappa < b_j^\eta \quad & \text{if } i<j \text{ or } i=j \text{ and } \kappa = 1,\ \eta=-1 \end{aligned}$$

This linear ordering induces the following (so called left-to-right lexicographic) ordering of the elements of $(A^\pm \cup B^\pm)^*$. Given $U \equiv x_1^{\epsilon_1}\cdots x_m^{\epsilon_m}$ and $V \equiv y_1^{\nu_1}\cdots y_n^{\nu_n}$ we say that U is less than V and write $U < V$, provided one of the following holds:

(i) $m < n$ and $x_i^{\epsilon_i} \equiv y_i^{\nu_i}$ for $1 \le i \le m$, or

(ii) there exist j, $1 \le j \le \min(m,n)$ such that $x_i^{\epsilon_i} \equiv y_i^{\nu_i}$ for $1 \le i < j$ and $x_j^{\epsilon_j} < y_j^{\nu_j}$.

The minimal element in $g_\Gamma(v_1, v_2)$, with respect to this ordering, is called a *C-geodesic* (see Def. 2.1.1).

Given $\mathcal{G} \in \mathcal{T}$, with its presentation $P_G = \langle A, B \mid \mathcal{R}_4, \mathcal{R}_2\rangle$, let $\mathcal{R}_{4*}$ be the symmetrization of $\mathcal{R}_4$ and let $\bar{\mathcal{R}}_{4*}$ be the set of all words which are $\mathcal{R}_2$-deducible from $\mathcal{R}_{4*}$. In other words, if $b^\epsilon a^\mu \beta^\nu \alpha^\iota$ is in $\mathcal{R}_{4*}$ and a^2 (or b^2) is also in $\mathcal{R}_2$, then both $b^\epsilon a^\mu \beta^\nu \alpha^\iota$ and $b^\epsilon a^{-\mu} \beta^\nu \alpha^\iota$ (or $b^{-\epsilon} a^\mu \beta^\nu \alpha^\iota$) are in $\bar{\mathcal{R}}_{4*}$, although $b^\epsilon a^{-\mu} \beta^\nu \alpha^\iota$ (or $b^{-\epsilon} a^\mu \beta^\nu \alpha^\iota$) is not a cyclic permutation of $b^\epsilon a^\mu \beta^\nu \alpha^\iota$ and does not belong to $\mathcal{R}_{4*}$. Note that $\mathcal{R}_{4*}$ and $\bar{\mathcal{R}}_{4*}$ are finite.

The set $\bar{\mathcal{R}}_{4*}$ ($\mathcal{R}_{4*}$) is a disjoint union of two sets, the set $B(\bar{\mathcal{R}}_{4*})$ ($B(\mathcal{R}_{4*})$) of all words in $\bar{\mathcal{R}}_{4*}$ ($\mathcal{R}_{4*}$) which begin with a B-symbol and the set $A(\bar{\mathcal{R}}_{4*}) = \bar{\mathcal{R}}_{4*} \setminus B(\bar{\mathcal{R}}_{4*})$. Similarly, the set $\mathcal{R}_{2*}$ is a disjoint union of two sets, the set $\mathcal{R}_2$ and the set $\mathcal{R}_2^- = \{r^{-1} \mid r \in \mathcal{R}_2\}$.

EXAMPLE 2.1.1. Let $P_G = \langle a, \alpha, b, \beta \mid ba\beta\alpha, a^2\rangle$ be a presentation of $G \cong \mathbb{Z}_2 * \mathbb{Z} * \mathbb{Z} \in \mathcal{T}$. Then

$$\begin{aligned} \bar{\mathcal{R}}_{4*} = \mathcal{R}_{4*} \cup \{\, & ba^{-1}\beta\alpha, a^{-1}\beta\alpha b, \beta\alpha b a^{-1}, \alpha b a^{-1}\beta, \\ & \alpha^{-1}\beta^{-1}ab^{-1}, b^{-1}\alpha^{-1}\beta^{-1}a, ab^{-1}\alpha^{-1}\beta^{-1}, \beta^{-1}ab^{-1}\alpha^{-1}\,\} \end{aligned}$$

We may notice that, beginning with the definition of our class $\mathcal{T}$, there is a certain asymmetry in "favor" of the B-symbols. We will continue to favor the B-symbols in the following definitions of "left side of a relation". We define

$$\mathcal{L}(\bar{\mathcal{R}}_{4*}) \stackrel{\text{def}}{=} \{\omega \mid \omega \text{ is a prefix of length two of a word in } B(\bar{\mathcal{R}}_{4*})\}$$

$$\mathcal{L}(\mathcal{R}_{4*}) \stackrel{\text{def}}{=} \{\omega \mid \omega \text{ is a prefix of length two of a word in } B(\mathcal{R}_{4*})\}$$

$$\mathcal{L}(\mathcal{R}_{2}) \stackrel{\text{def}}{=} \{\xi \mid \xi \text{ is a prefix of length one of a word in } \mathcal{R}_2^-\}$$

To illustrate this, in the previous example we have $\mathcal{L}(\bar{\mathcal{R}}_{4*}) = \{ba, \beta\alpha, \beta^{-1}a, \beta^{-1}a^{-1}, b^{-1}\alpha^{-1}, ba^{-1}\}$ and $\mathcal{L}(\mathcal{R}_2) = \{a^{-1}\}$.

DEFINITION 2.1.1. A freely reduced word $W \equiv \xi_1^{\epsilon_1} \dots \xi_n^{\epsilon_n}$ ($\xi_i \in A \cup B$ and $\epsilon_i = \pm 1$) is a *C-geodesic* if

(1) $\xi_i^{\epsilon_i} \notin \mathcal{L}(\mathcal{R}_2)$ for all $1 \le i \le n$.
(2) $\xi_i^{\epsilon_i}\xi_{i+1}^{\epsilon_{i+1}} \notin \mathcal{L}(\bar{\mathcal{R}}_{4*}) \cup \mathcal{R}_2$ for all $1 \le i \le n-1$.

EXAMPLE 2.1.2. Let G be given by the following presentation:

$$P_G = \langle a, \alpha, b, \beta \mid ba\beta^{-1}\alpha, a^2, \alpha^2, b^2 \rangle$$

We can easily see that $G \cong \mathbb{Z}_2 * \mathbb{Z}_2 * \mathbb{Z}_2$. We have

$$\mathcal{L}(\bar{\mathcal{R}}_{4*}) = \{\, ba, ba^{-1}, b^{-1}\alpha^{-1}, b^{-1}\alpha, b\alpha^{-1}, b\alpha, \beta a^{-1}, \beta a, \beta^{-1}\alpha, \beta^{-1}\alpha^{-1} \,\}$$
$$\mathcal{L}(\mathcal{R}_2) = \{\, a^{-1}, \alpha^{-1}, b^{-1} \,\}.$$

The word $b\alpha\beta$ is not a C-geodesic because $b\alpha \in \mathcal{L}(\bar{\mathcal{R}}_{4*})$, but the words $b\beta b$, $a\alpha b$ or the empty word are examples of C-geodesics.

Interpreting the words in $A^{\pm} \cup B^{\pm}$ as paths in the Cayley graph $\Gamma(G, A \cup B)$ for $G \in \mathcal{T}$ we can see that C-geodesics correspond to these paths in $\Gamma(G, A \cup B)$ which satisfy the following property:

Whenever a C-geodesic path $W(t)$ passes through two opposite vertices in a 4-cell, first it will pass the edge labeled with an A-symbol and then the edge labeled with a B-symbol. If a $W(t)$ passes through two opposite vertices in a 2-cell, it will pass through the edge labeled x for $x \in A \cup B$.

PROPOSITION 2.1.1. *Let $W \equiv \xi_1^{\epsilon_1} \dots \xi_n^{\epsilon_n}$ be a C-geodesic representing an element $g \in G$, $G \in \mathcal{T}$ and (M, ϕ, p, v_0) be a singular disk diagram which includes W. If $W(t)$ passes through two consecutive edges e and e' of $\dot{M}^{(1)}$ such that $\phi(e) \in B^{\pm}$, and $\phi(e') \in A^{\pm}$, then there is no 4-cell $B \in M^{(2)}$ having e and e' among its boundary edges.* □

Note that a prefix of a C-geodesic is again a C-geodesic. Also a suffix of a C-geodesic is C-geodesic. A simple example can show that the concatenation of two C-geodesics is not necessarily a C-geodesic.

How do we obtain a C-geodesic representative of an element $g \in G$? Given W as a geodesic representative of g, using the defining relators in P_G, we will "push" the B-symbols in W to the right, whenever it is possible.

PROPOSITION 2.1.2. *Every element $g \in G$, $G \in \mathcal{T}$, has a C-geodesic representative.*

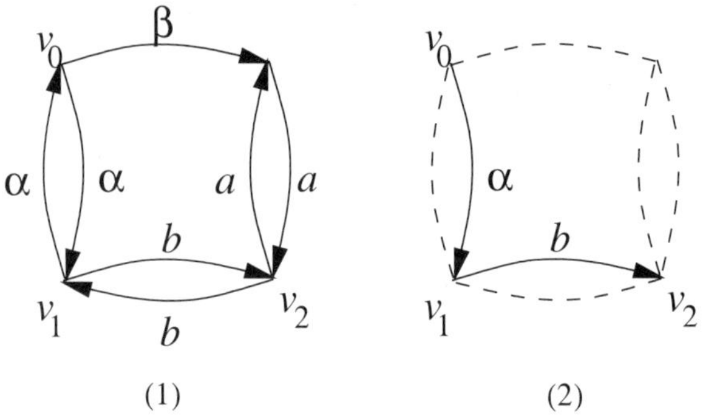

FIGURE 2.1. The figure on the left shows a disk diagram (M, ϕ, p, v_0) for a freely reduced relator $\alpha\beta ab$. If v_1 represents a vertex in $\Gamma(\mathbb{Z}_2 * \mathbb{Z}_2 * \mathbb{Z}_2, \{a, \alpha, b, \beta\})$ then there is only one C-geodesic having v_1 as its initial and v_2 as its terminal vertex, as indicated in figure (2).

PROOF. (Compare [4, p. 318]) Let $W \equiv \xi_1^{\epsilon_1} \dots \xi_n^{\epsilon_n}$, $\xi_i \in A \cup B$, $\epsilon_i = \pm 1$ be a geodesic representative of $g \in G$ and $P_G = \langle A, B \mid \mathcal{R}_4, \mathcal{R}_2 \rangle$ be a presentation of $G \in \mathcal{T}$. We will say that the symbols b^{ϵ_b} and a^{ϵ_a} in W form a (B, A) pair if there exist subwords W_1, W_2 and W_3 of W such that

$$W \equiv W_1 b^{\epsilon_b} W_2 a^{\epsilon_a} W_3.$$

Let $H(W)$ be the number of (B, A) pairs in W. (For example if $W \equiv b^{-1} a\alpha b\beta\alpha^{-1}$ then $H(W) = 5$.) If $\xi_j^{\epsilon_j} \xi_{j+1}^{\epsilon_{j+1}}$, $1 \leq j \leq n-1$, is a subword of W which is an element of $\mathcal{L}(\bar{\mathcal{R}}_{4*})$ (i.e., $\xi_j^{\epsilon_j}$ and $\xi_{j+1}^{\epsilon_{j+1}}$ form a (B, A) pair) we can replace it with $\eta_j^{\nu_j} \eta_{j+1}^{\nu_{j+1}}$ where $\xi_j^{\epsilon_j} \xi_{j+1}^{\epsilon_{j+1}} \eta_{j+1}^{-\nu_{j+1}} \eta_j^{-\nu_j}$ is a relator in $\bar{\mathcal{R}}_{4*}$. Note that η_j and η_{j+1} are unique, but the values of ν_j and ν_{j+1} depend on the set $\mathcal{R}_2$. We will obtain a word $W' \equiv \xi_1^{\epsilon_1} \dots \xi_{j-1}^{\epsilon_{j-1}} \eta_j^{\nu_j} \eta_{j+1}^{\nu_{j+1}} \xi_{j+2}^{\epsilon_{j+2}} \dots \xi_n^{\epsilon_n}$ which is also a representative of g having the same length as W and is freely reduced. For, if W' was not freely reduced, say if $\xi_{j-1} \equiv \eta_j$ and $\epsilon_{j-1} + \nu_j = 0$, then W would contain a subword $a^{\epsilon_a} b^{\epsilon_b} \alpha^{\epsilon_\alpha}$ with $b^{\epsilon_b} \alpha^{\epsilon_\alpha} \equiv \xi_j^{\epsilon_j} \xi_{j+1}^{\epsilon_{j+1}}$, $a^{\epsilon_a} \equiv \xi_{j-1}^{\epsilon_{j-1}}$ and $\beta \in B$ such that $b^{\epsilon_b} \alpha^{\epsilon_\alpha} \beta^{-\epsilon_\beta} a^{\epsilon_a} \in \bar{\mathcal{R}}_{4*}$. So, a shorter representative of g would exist, namely $\xi_1^{\epsilon_1} \dots \xi_{j-2}^{\epsilon_{j-2}} \beta^{\epsilon_\beta} \xi_{j+2}^{\epsilon_{j+2}} \dots \xi_n^{\epsilon_n}$. Similarly, we can show that either $\eta_{j+1} \not\equiv \xi_{j+2}$ or if $\eta_{j+1} \equiv \xi_{j+2}$ then $\epsilon_{j+2} \neq -\eta_{j+1}$.

For the new representative W' of g we have that $H(W') < H(W)$. Continuing in this way we will end up with a word $W^{(k)}$ which either does not contain (B, A) pairs, or does not contain subwords of length 2 which are elements of $\mathcal{L}(\bar{\mathcal{R}}_{4*})$. So $W^{(k)}$ will satisfy (2) of Definition 2.1.1. Now, if $W^{(k)} \equiv \mu_1^{\lambda_1} \dots \mu_i^{-1} \dots \mu_n^{\lambda_n}$ and $\mu_i^2 \in \mathcal{R}_2$, we will replace μ_i^{-1} by μ_i for all such i and obtain a C-geodesic representative of g. □

DEFINITION 2.1.2. Let D be a disk diagram over a $C(2,4) - T(4)$ presentation P_G of some group G. A *chain* in D is a connected segment of the boundary of D which begins and ends at vertices with d_4-degree equal to 1 and all of whose vertices in between (if any) have d_4-degree equal to 2 (see Figure 2.2 (1)).

Note that if a disk diagram (D, ϕ, p, v_0) includes a geodesic W on its boundary and $\phi(p_c)$ is the label of a chain p_c in D, such that $\phi(p_c)$ is a subword of W, then $\phi(p_c)$ is not a proper subword of W.

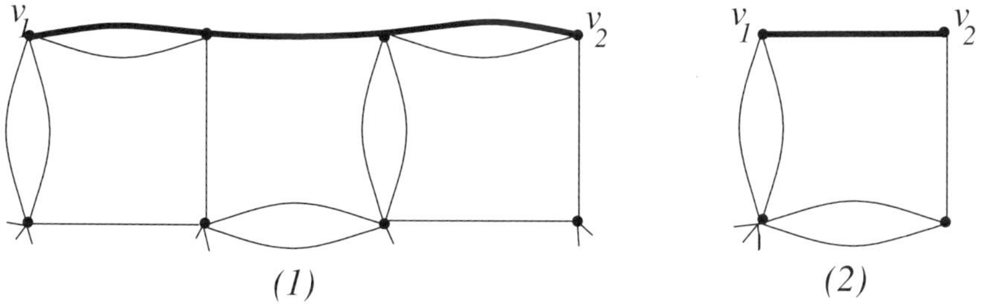

FIGURE 2.2

If a chain doesn't contain a vertex v such that $d_4(v) = 2$ we will call it *degenerate* (see Figure 2.2 (2)).

Our next step is to characterize (in terms of the number of chains) disk diagrams for $C(2,4)-T(4)$ presentations P_G for $G \in \mathcal{T}$. If D is a disk diagram which includes the word W, let the number of vertices $|D^{(0)}|$ of D, edges $|D^{(1)}|$ and cells $|D^{(2)}|$ be labeled as V, E and F respectively. Let $\sum_v$, $\overset{\bullet}{\sum_v}$ and $\overset{\circ}{\sum_v}$ stand for summation over all vertices in D, all boundary vertices in D and all interior vertices in D respectively. Let $\sum_B$ stand for summation over all cells in D. The sets of all 2-cells and 4-cells in D will be labeled with $\mathcal{B}_2$ and $\mathcal{B}_4$ respectively. Of course, $D^{(2)} = \mathcal{B}_2 \cup \mathcal{B}_4$ and $|D^{(2)}| = |\mathcal{B}_2| + |\mathcal{B}_4|$. Using the formula for the Euler characteristic of D we have the following lemma.

LEMMA 2.1.1. (comp. [13]) $\overset{\bullet}{\sum_v}[3 - d(v)] + \overset{\circ}{\sum_v}[4 - d(v)] + 2|\mathcal{B}_2| = 4$

PROOF. We will start with the Euler characteristic for a disk

$$V - E + F = 1.$$

We also have that

$$(*) \qquad \begin{aligned} \sum_v d(v) &= 2E \\ \sum_B d(B) + d(D) &= 2E \end{aligned}$$

Eliminating E from the equations in $(*)$ we will get

$$\sum_v d(v) - \sum_B d(B) - d(D) = 0$$

Using this and $4V - 4E + 4F = 4$, with $\sum_v d(v) = 2E$ we will get

$$4V - 2\sum_v d(v) + 4F = 4 \qquad \text{or}$$

$$4V - \sum_v d(v) - d(D) - \sum_B d(B) + 4F = 4$$

Finally

$$\sum_v [4 - d(v)] - d(D) + \sum_B [4 - d(B)] = 4.$$

If we decompose the first sum into a summation $\overset{\bullet}{\sum}$ over all boundary vertices in D and summation $\overset{\circ}{\sum}$ over all interior vertices in D, using the fact that $d(D) = \overset{\bullet}{\sum_v} 1$, we will get

$$\overset{\bullet}{\sum_v} [3 - d(v)] + \overset{\circ}{\sum_v} [4 - d(v)] + \sum_B [4 - d(B)] = 4$$

But $\sum_B = \sum_{\mathcal{B}_2} + \sum_{\mathcal{B}_4}$ and for all $B \in \mathcal{B}_4$, $d(B) = 4$. Thus

(2.1.1) $$\overset{\bullet}{\sum_v} [3 - d(v)] + \overset{\circ}{\sum_v} [4 - d(v)] + 2|\mathcal{B}_2| = 4.$$

□

In the following three lemmas we will simplify formula (2.1.1).

LEMMA 2.1.2. *Let (M, ϕ, p, v_0) be a singular disk diagram over P_G, $G \in \mathcal{T}$, and let B_2 be a 2-cell in $M^{(2)}$ i.e. $\phi(\dot{B}_2) \in \mathcal{R}_{2*}$. Then $\bar{B}_2$, the closure of B_2 in $\mathbb{E}$, is topologically a disk.*

PROOF. Let e and e' be the boundary edges of B_2 and let v_0' and v_0'' be their end vertices. We want to show that $v_0' \neq v_0''$. Assume that $v_0' = v_0''$. Then each of the boundary edges e and e' is a loop, i.e., a simple reduced closed path of length one. Because M is simply connected each of the edges e and e' will bound a disk in $\mathbb{E}$. Take e to bound a disk which is a subset of the disk bounded by e' and let (M', ϕ, e, v_0') be a subdiagram of M with a boundary cycle consisting of only one edge e (see Figure 2.3(1)).

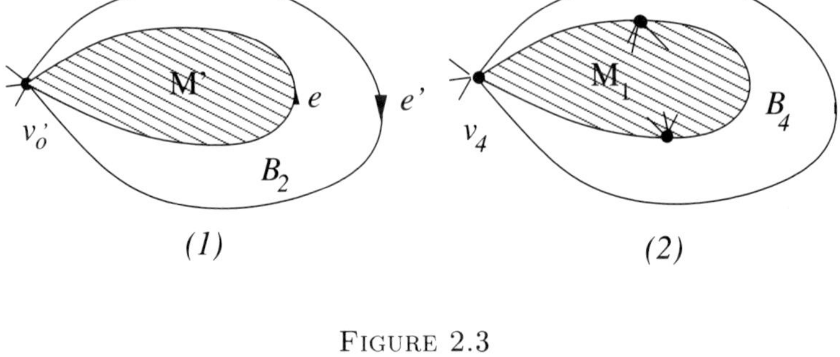

FIGURE 2.3

Because (M, ϕ, p, v_0) is reduced, e is not a common edge of B_2 and some other 2-cell in $M^{(2)}$, therefore M' has at least one (interior) 4-cell and the degree of v_0', as a vertex in M', is ≥ 3. Since M has finitely many cells, we can assume that the closure of every 2-cell in M' is topologically a disk. Note that if v is an interior vertex in M' incident to a 2-cell then $d(v) \geq 5$, so

$$\overset{\circ}{\sum_v} [4 - d(v)] + 2|\mathcal{B}_2| \leq 0$$

i.e., the left side of the formula (2.1.1) is ≤ 0. Consequently, no such map M' exists. □

LEMMA 2.1.3. *Let (M, ϕ, p, v_0) be a singular disk diagram over P_G, $G \in \mathcal{T}$, and let B_4 be a 4-cell in $M^{(2)}$, i.e., $\phi(\dot{B}_4) \in \mathcal{R}_{4*}$. Then $\bar{B}_4$, the closure of B_4 in $\mathbb{E}$, is topologically a disk.*

PROOF. Assume that two of the vertices in B_4 coincide such that three of the edges of $\dot{B}_4$ form a simple loop p_1, beginning and ending at the same vertex ($= v_4$) (see Figure 2.3(2)).

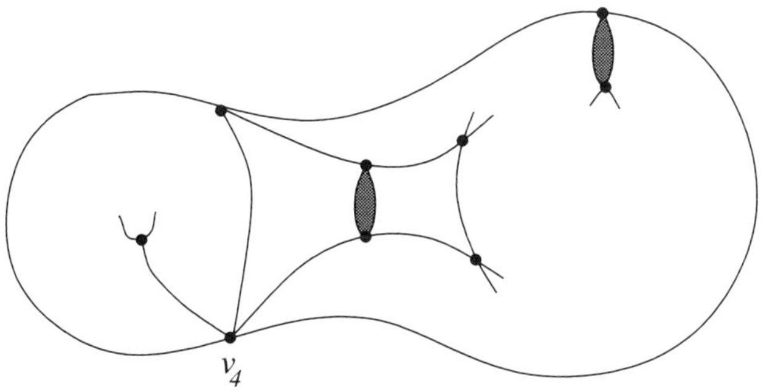

FIGURE 2.4

Because M is simply connected, p_1 will bound a disk, therefore, there is a subdiagram (M_1, ϕ, p_1, v_4) with three vertices on its boundary. Since M has finitely many cells, we can choose B_4 such that the closure of every 4-cell in M_1 is topologically a disk and for all $v \in \dot{M}_1^{(0)}$, $d(v) \geq 3$. We can also assume that none of the edges of p_1 is a boundary edge of a 2-cell.

Assume that i $(0 \leq i \leq 3)$, is the number of boundary vertices in M_1 incident to 2-cells. Then

$$\sum_{v \in \dot{M}_1} [3 - d(v)] \leq -i.$$

Before applying formula (2.1.1) to M_1, we will represent the set $\mathcal{B}_2$ as a disjoint union of the sets $\mathring{\mathcal{B}}_2$, consisting of all 2-cells with no vertex on the boundary of M_1, and the set $\dot{\mathcal{B}}_2$ consisting of all 2-cells with exactly one of their vertices on the boundary of M_1. Note that if there is a 2-cell in M_1 with both of its vertices in $\mathring{M}_1^{(0)}$, then there exist a submap M_1' of M_1 with only two boundary vertices and two boundary edges such that the left side of (2.1.1) is ≤ 0. Similarly, the set $\mathring{M}_1^{(0)}$ of interior vertices of M_1, is disjoint union of the following two sets:

(1) $M_{10} = \{v \in \mathring{M}_1^{(0)} \mid v \text{ is incident to a 2-cell with none of its vertices in } \dot{M}_1^{(0)}\}$

(2) $M_{11} = \{v \in \mathring{M}_1^{(0)} \mid v \text{ is incident to a 2-cell with exactly one of its vertices in } \dot{M}_1^{(0)}\}$

See Figure 2.4 for illustration. Obviously

$$\sum_{v\in M_1}^{\circ} = \sum_{v\in M_{10}}^{\circ} + \sum_{v\in M_{11}}^{\circ}.$$

Finally, computing the left side of formula (2.1.1) for M_1 we will get

$$\sum_{v\in M_1}^{\cdot} [3 - d(v)] + \sum_{v\in M_{11}}^{\circ} [4 - d(v)] + 2\left|\dot{\mathcal{B}}_2\right| \leq 0$$

and

$$\sum_{v\in M_{10}}^{\circ} [4 - d(v)] + 2\left|\overset{\circ}{\mathcal{B}}_2\right| \leq 0.$$

Therefore, the left side of the formula (2.1.1) is ≤ 0. Consequently, no such map M_1 exists. Similar arguments apply in the case when the path p_1 has two edges or only one edge. □

Using the last two lemmas one can easily prove the following:

LEMMA 2.1.4. *If M is a disk diagram over P_G, $G \in \mathcal{T}$ and $v \in M^{(0)}$, then*

$$v \in \dot{M}^{(0)} \Rightarrow d(v) = d_2(v) + d_4(v) + 1$$

and

$$v \in \overset{\circ}{M}^{(0)} \Rightarrow d(v) = d_2(v) + d_4(v). \quad \square$$

If we take into account the connection between d, d_2 and d_4 as a consequence of (2.1.1) we will get

$$\sum_{v}^{\bullet}[3 - (d_2(v) + d_4(v) + 1)] + \sum_{v}^{\circ}[4 - (d_2(v) + d_4(v))] + 2|\mathcal{B}_2| = 4,$$

or

$$\sum_{v}^{\bullet}[2 - d_4(v)] - \sum_{v}^{\bullet} d_2(v) + \sum_{v}^{\circ}[4 - d_4(v)] - \sum_{v}^{\circ} d_2(v) + 2|\mathcal{B}_2| = 4$$

Obviously, $2|\mathcal{B}_2| = \sum_{v}^{\bullet} d_2(v) + \sum_{v}^{\circ} d_2(v)$. Therefore the last formula can be written as

$$\sum_{v}^{\bullet}[2 - d_4(v)] + \sum_{v}^{\circ}[4 - d_4(v)] = 4. \tag{2.1.2}$$

This formula will play a crucial role in determining the "shape" of geodesic triangles and digons in $\Gamma(G, A \cup B)$ for $G \in \mathcal{T}$.

2.2. Shape of Geodesic Triangles and Digons for Groups in $\mathcal{T}$

Before characterizing the structure of the geodesic triangles and digons in $\Gamma(G, A \cup B)$ of $C(2,4) - T(4)$ groups $G \in \mathcal{T}$, we derive a few consequences of the formulas (2.1.1) and (2.1.2).

COROLLARY 2.2.1. *Let D be a disk diagram over P_G, $G \in \mathcal{T}$ with at least one 4-cell. Then*

$$\left|\{v \mid v \in \dot{D}^{(0)}, d_4(v) = 1\}\right| \geq 4 + \left|\{v \mid v \in \dot{D}^{(0)}, d_4(v) \geq 3\}\right|$$

PROOF. For the interior vertices in D, we have that

$$\mathop{\mathring{\sum}}_{v} [4 - d_4(v)] \leq 0.$$

Using this and (2.1.2), it follows that

$$\mathop{\dot{\sum}}_{v} [2 - d_4(v)] \geq 4.$$

Expanding the last inequality we obtain

$$\begin{aligned}\left|\{v \mid d_4(v) = 1, v \in \dot{D}^{(0)}\}\right| \geq 4 &+ \left|\{v \mid d_4(v) = 3, v \in \dot{D}^{(0)}\}\right| \\ &+ 2\left|\{v \mid d_4(v) = 4, v \in \dot{D}^{(0)}\}\right| + \dots\end{aligned}$$

□

In words, on the boundary of a disk diagram over P_G, $G \in \mathcal{T}$, we have at least four more vertices of 4-degree one, then there are vertices of 4-degree greater than or equal to three. The vertices of 4-degree greater or equal to three will divide the boundary of D into segments such that at least four of these segments will contain no interior vertex of 4-degree greater than or equal to three.

COROLLARY 2.2.2. *On the boundary of a disk diagram over a presentation P_G, $G \in \mathcal{T}$, with at least one 4-cell, there are at least four chains with disjoint interiors.* □

The next corollary will throw more light on the "BABA-structure" of the relators over P_G.

COROLLARY 2.2.3. *Let (M, ϕ, p, v_0) be a singular disk diagram with at least one 4-cell. If $W \equiv \phi(p)$, then there are nonempty A-words A_1, A_1', A_2 and non-empty B-words B_1, B_2 and B_2' and words W_1, W_1', W_2 and W_2' such that*

$$W \equiv W_1 A_1 B_1 A_1' W_1'$$

or

$$W \equiv W_2 B_2 A_2 B_2' W_2'$$

□

Following the proof of Proposition 2.1.2, we see that starting with a geodesic representative of $g \in G$, we can transform it into a C-geodesic representative. This, of course, does not prove that C-geodesics are geodesics, because a C-geodesic doesn't have to be obtained by this construction. Thus, we must prove the following proposition:

THEOREM 2.2.1. *Every C-geodesic representative of an element $g \in G$, $G \in \mathcal{T}$, is also a geodesic representative of the same element.*

PROOF. Let W be a C-geodesic and U a geodesic representative of the same element $g \in G$. We will show that they have the same length, i.e., $|W| = |U|$.

First, we will show that the only C-geodesic representative of $1 \in G$ is the empty word.

Let W be the C-geodesic representative of 1, i.e., $\bar{W} = 1$. We will induct on $|W|$. If $|W| = 1$ then there will be a disk diagram D with only one edge on its boundary. But this contradicts the formula (2.1.2), so W must be the empty word. Let the proposition be true for all C-geodesics W such that $|W| < k$ and assume that $|W| = k$. There will be a singular disk diagram (M, ϕ, p, v_0) such that $\phi(p) \equiv W$. If M has a cut vertex, then there will be a subdiagram (M', ϕ, p', v') of M such that $v' \in \dot{M}^{(0)}$ and $\phi(p')$ is a subword of W. By the inductive hypothesis, $\phi(p')$ must be the empty word and the conclusion will follow. So we can assume that (M, ϕ, p, v_0) has no cut vertices.

If M contains at least one 4-cell, by Corollary 2.2.2, there are at least four chains with disjoint interiors on the boundary of M, so W fails to be a C-geodesic. Otherwise, $M^{(2)}$ will contain only one 2-cell such that $\phi(p) \in \mathcal{R}_{2*}$, i.e., W will fail to be a C-geodesic.

So we can assume that $g \neq 1$ and we will continue with the proof of the proposition by inducting on the sum of the lengths $|U| + |W|$ of U and W.

By Proposition 2.1.2, every geodesic can be transformed to a C-geodesic, so we can assume that U is a C-geodesic too. If $|U| + |W| = 2$ then $|W| = |U| = 1$. If $U \not\equiv W$ there will be a disk diagram (D, ϕ, p, v_0) such that $\phi(p) = UW^{-1}$ and $\dot{D}^0$ has only two elements. Again a contradiction with the formula (2.1.2), so $U \equiv W$.

Assume that the statement is true for all U and W such that $|U| + |W| < k$ and let $|U| + |W| = k$. Let (M, ϕ, p, v_0) be a singular disk diagram for the relator UW^{-1}. If M has a cut vertex v_1 it will consist of two singular disk diagrams (M', ϕ, p', v_0) and (M'', ϕ, p'', v_1) joined at v_1. In this case the result follows by the inductive hypothesis since the initial and the terminal segments of C-geodesics are C-geodesics. So, we can assume that M is a disk diagram. If M contains at least one 4-cell, by Corollary 2.2.2, there will be at least four chains with disjoint interiors. None of these can be entirely contained in W or U. So M must contain at most four chains, two of which share the vertex v_0 and the other two some other vertex v' (see Figure 2.5 (1)). If v_0 is a common vertex of two chains, one of them must have a label some nonempty B-word, so one of U or W will fail to be a C-geodesic. It remains to consider the case when M does not contain a 4-cell. Because M is reduced, it must be composed of only one 2-cell (see Figure 2.5 (2)). Again we have the case $|W| = |U| = 1$, i.e., $W \equiv U$. □

In fact, we have shown even more:

COROLLARY 2.2.4. *For every element $g \in G$, there is a unique C-geodesic W in the Cayley graph $\Gamma(G, A \cup B)$, $G \in \mathcal{T}$ which represents g.* □

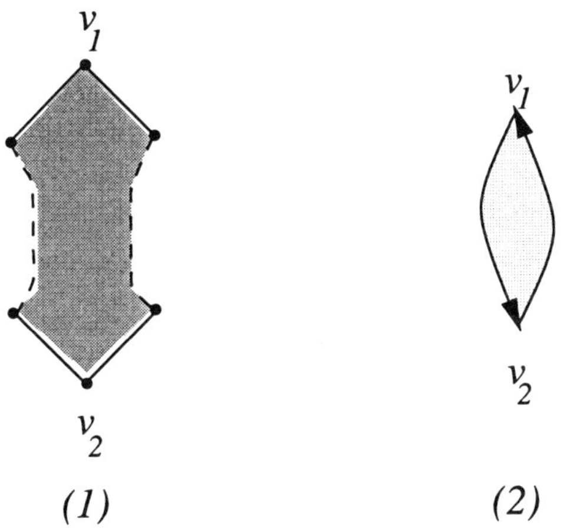

FIGURE 2.5

COROLLARY 2.2.5. *If W is a C-geodesic which is a B-word (A-word), then there is no singular disk diagram (M, ϕ, p, v_0) such that $\phi(p)$ is a B-word (A-word) with W being a subword of $\phi(p)$.*

PROOF. This follows from Corollary 2.2.4 and Corollary 2.2.3. □

LEMMA 2.2.1. *Let W be a geodesic such that $|W| \geq 2$ and let (D_W, ϕ, p, v_0) be a disk diagram which includes W. Then*

$$\sum_{v \in Int(W)} [2 - d_4(v)] \leq 1.$$

PROOF. We know that D_W has at least four chains on its boundary. W cannot contain a chain in its interior, so W can have at most two chains both on its ends. The inequality then follows easily. □

PROPOSITION 2.2.1. *Let U and W be two geodesics representing the same element $g \in G$ and let D be a disk diagram over P_G for the relator UW^{-1}. Then there is no interior vertex $v \in \overset{\circ}{D}^{(0)}$ such that $d_4(v) > 4$.*

PROOF. Since D includes U and W, by Lemma 2.2.1 we have that:

$$\sum_{v \in Int(U)} [2 - d_4(v)] \leq 1 \quad \text{and} \quad \sum_{v \in Int(W)} [2 - d_4(v)] \leq 1.$$

Following the proof of Proposition 1.2.1 we can see that there is no interior vertex v in D, such that $d_4(v) = 3$. Similarly, we can show that no interior vertex in D has d_4-degree equal to 5, therefore, if there was a vertex $v \in \overset{\circ}{D}^{(0)}$ such that $d_4(v) > 4$ then $d_4(v) \geq 6$ and

$$\sum_{v \in \overset{\circ}{D}^{(0)}} [4 - d_4(v)] \leq -2.$$

In this case we would get

$$\sum_v^{\bullet}[2-d_4(v)]+\sum_v^{\circ}[4-d_4(v)]$$

$$\leq 2+\sum_{v\in Int(U)\cup Int(W)}[2-d_4(v)]+\sum_v^{\circ}[4-d_4(v)]\leq 2$$

which contradicts (2.1.2). □

COROLLARY 2.2.6. *Let U and W be two geodesics representing the same element $g\in G$ and let D be a disk diagram over P_G for the relutor UW^{-1}. Then there is no boundary vertex $\dot{v}$ in D such that $d_4(v)>3$.*

PROOF. By Proposition 2.2.1 if v is an interior vertex in D then $d_4(v)=4$. It follows from (2.1.2) that

(1) $$\sum_v^{\bullet}[2-d_4(v)]=4$$

We know that U and W must begin and end with a chain (otherwise they will contain a chain in their interior and would fail to be geodesics). Hence, D has at most four chains. Let v_1 and v_2 be the end vertices of the beginning chains in U and W respectively, v_0 their common initial vertex and v_3 the common vertex of the terminal chains of U and W respectively (see Figure 2.6).

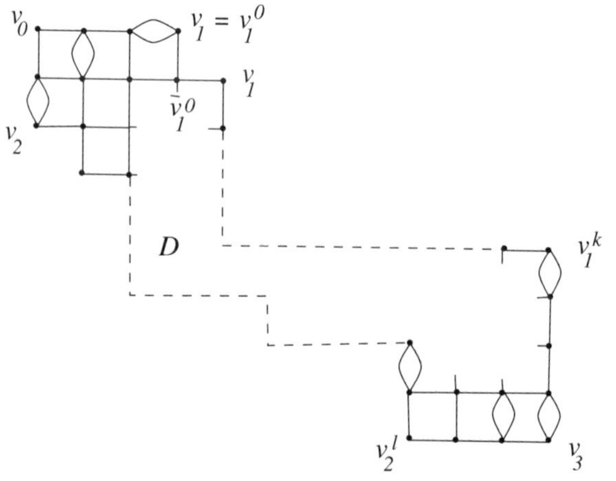

FIGURE 2.6

Using (2.2.1) we have

(2) $$\sum_{v\neq v_0,v_1,v_2,v_3}^{\bullet}[2-d_4(v)]=0.$$

Let $v_1^0=v_1, v_1^1,\ldots,v_1^k$, $(v_1^k\neq v_3)$, $k\geq 0$ be the vertices on $U(t)$ with d_4-degree equal to 1 and $v_2^0=v_2, v_2^1,\ldots,v_2^l$, $(v_2^l\neq v_3)$, $l\geq 0$, be the vertices on $W(t)$ with d_4-degree equal to 1. For every i, $0\leq i\leq k-1$, and every j, $0\leq j\leq l-1$, there will be a vertex $\bar{v}_1^i$ ($\bar{v}_2^j$) between the vertices v_1^i and v_1^{i+1} (v_2^j and v_2^{j+1}) with d_4-degree greater than or equal to 3. Otherwise, v_1^i and v_1^{i+1} (v_2^j and v_2^{j+1}) will be the end vertices of a chain contained in U (W). If there exists a vertex v on the boundary of D such that $d_4(v) \geq 4$, then we can choose v to be one of $\bar{v}_1^i$, $0 \leq i \leq k-1$, or $\bar{v}_2^j$, $0 \leq j \leq l-1$. Now, if we sum in (2.2.2) over all pairs $(\bar{v}_1^i, v_1^{i+1})$ and $\left(\bar{v}_2^j, v_2^{j+1}\right)$ we will get

$$(3) \qquad \dot{\sum_{v \neq v_0, v_1, v_2, v_3}} [2 - d_4(v)] \leq -1$$

which contradicts (2.2.2). □

Using induction on the number of disk components in a singular disk diagram (M, ϕ, p, v_0) we have the following corollary.

COROLLARY 2.2.7. *Let (M, ϕ, p, v_0) be a singular disk diagram which includes two geodesics U and W such that $\phi(p) \equiv UW^{-1}$. Then no interior vertex v in M has $d_4(v) > 4$ and no boundary vertex v' in M has $d_4(v') > 3$.* □

This corollary is a motivation for the next definition of a rectangular disk diagram (RDD) and almost rectangular disk diagram (ARDD).

DEFINITION 2.2.1. *An almost rectangular disk diagram* (M, ϕ, p, v_0) is a disk diagram in which every interior vertex v has $d_4(v) = 4$ and every boundary vertex has d_4-degree less than or equal to three. *A rectangular disk diagram* is an almost rectangular disk diagram in which there are no boundary vertices with d_4-degree equal to 3.

An almost rectangular disk diagram is called *regular* if it has only four chains and every chain labeled with some A-word (B-word) has a common vertex with a chain labeled with some B-word (A-word). Obviously the "opposite" chains, i.e., the chains which do not have a common vertex, have labels A-words or B-words. Common vertices to A-chains and B-chains are called *corners* of that regular ARDD. Two corners are called *opposite* if there is no chain that contains both of them. Note that a singular disk diagram (M, ϕ, p, v_0) with no 4-cells is ARDD by definition. So, if not otherwise stated, it is assumed that an ARDD has at least one 4-cell.

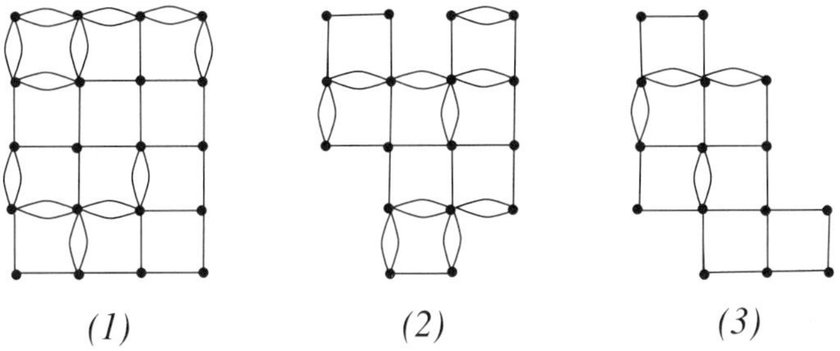

FIGURE 2.7. (1) is an example of RDD, (2) is an example of ARDD which is not a regular ARDD and (3) is an example of a regular ARDD which is not RDD. For simplicity, we have omitted the labels and the orientation of the edges.

COROLLARY 2.2.8. *Let U and W be two geodesics representing the same element $g \in G$ and let D be a disk component in a singular disk diagram over P_G for the relator UW^{-1}. Then D is a regular ARDD.*

PROOF. Clearly D is an ARDD and if D contains more than four chins, one of the geodesics U or W will contain a chain in its interior, which is a contradiction. We reach a similar contradiction if we assume that one of the chains of the boundary of D has no common vertex with any other chain of the boundary of D. Therefore D must be a regular ARDD. □

The last corollary can be used to obtain **all** singular disk diagrams (M, ϕ, p, v_0) which include two geodesics U and W representing the same element $g \in G$.

THEOREM 2.2.2. *Let $U \equiv a_1^{\epsilon_1} \dots a_k^{\epsilon_k} b_{k+1}^{\epsilon_{k+1}} \dots b_{k+l}^{\epsilon_{k+l}}$, $\epsilon_i = \pm 1$, $k, l \geq 1$ be a C-geodesic representing the element $g = \bar{U} \in G$, $G \in \mathcal{T}$. Then there is a unique singular disk diagram $(D_{[U]}, \phi, pqs_U^{-1}, v_0)$, with $\phi(p) \equiv a_1^{\epsilon_1} \dots a_k^{\epsilon_k}$ and $\phi(q) \equiv b_{k+1}^{\epsilon_{k+1}} \dots b_{k+l}^{\epsilon_{k+l}}$ such that:*

1. *Either no disk component of $D_{[U]}$ contains a 4-cell; or exactly one contains a 4-cell, in which case that disk component is an ARDD.*
2. *There is a vertex v_1 on the boundary of $D_{[U]}$ such that for the boundary paths pq and s_U from v_0 to v_1, $\phi(pq) \equiv U$ and $U[\phi(s_U)]^{-1}$ is a relator over P_G.*
3. *$\phi(s_U^{-1})$ is a C-geodesic representing the element g^{-1} in G.*
4. *There exists a unique singular disk diagram $(\bar{D}_{[U]}, \phi, pq\bar{s}_U^{-1}, v_0)$ obtained from $D_{[U]}$ by attaching finitely many 2-cells along the path s_U, such that every geodesic representative of g will appear as a label on some path in $\bar{D}_{[U]}$ from v_0 to v_i.*
5. *Every other singular disk diagram which includes two geodesics on its boundary, representing the same element $g \in G$ is a subdiagram of $(\bar{D}_{[U]}, \phi, pq\bar{s}_U^{-1}, v_0)$.*

PROOF. Let V and V' be two geodesics representing the same element $g \in G$, $G \in \mathcal{T}$. We can rephrase the ordering on the set $g_\Gamma(1, g)$ of all geodesics representing the element g given in 2.1 by the following:

$$V \leq V' \text{ iff either } V \equiv V' \text{ or } V \equiv V_1 a^{\epsilon_a} V_2 \text{ and } V \equiv V_1 b^{\epsilon_b} V_2' \text{ or}$$
$$V \equiv V_1 x V_2 \text{ and } V' \equiv V_1 x^{-1} V_2' \text{ where } x \in A \cup B.$$

Note that by Corollary 2.2.8 the case $V \equiv V_1 a^{\epsilon_a} V_2$ and $V' \equiv V_1 \alpha^{\epsilon_a} V_2'$, $a, \alpha \in A$ or $V \equiv V_1 b^{\epsilon_b} V_2$ and $V' \equiv V_1 \beta^{\epsilon_\beta} V_2'$, $b, \beta \in B$ does not occur. Thus, given a singular disk diagram M which is also a regular ARDD, we can place M in the plane such that all edges labeled with A-symbols will be "vertical" and all edges labeled with B-symbols will be horizontal. First we construct the singular disk diagram $\bar{D}_{[U]}$ inductively and then we obtain $D_{[U]}$ from $\bar{D}_{[U]}$ by deleting finitely many 2-cells on the boundary of $\bar{D}_{[U]}$.

First let $M^{(0)}$ be a rectangular net of vertices in the plane defined by

$$M^{(0)} = \{(x, y) \mid x, y \in \mathbb{Z}, 0 \leq x \leq l, 0 \leq y \leq k\}.$$

We realize U as the edge path pq having $\{(0, i) \mid 0 \leq i \leq k\}$ as the set of vertices for p and $\{(j, k) \mid 1 \leq j \leq l\}$ as the set of vertices for q. Let $v_0 = (0, 0)$ and $v_1 = (k, l)$. We define $\bar{D}_0$ to be this edge path and this will be a singular disk diagram with no cells. If none of the a_i's or b_j's in U appears as a symbolin $\mathcal{R}_2$ and if $b_{k+1}^{-\epsilon_{k+1}} a_k^{-\epsilon_k} \notin \mathcal{L}(\mathcal{R}_{4*})$ we will take $D_{[U]}(= \bar{D}_{[U]})$ to be $\bar{D}_0$. By Corollary 2.2.4, $D_{[U]}$ satisfies (1)–(5).

Now, assume that $\bar{D}_n = (\bar{D}_n, \phi, pqs_n^{-1}, v_0)$, $n \geq 0$ has been constructed and it satisfies the following properties:

(1) $\bar{D}_n$ has at most one disk component with at least one 4-cell and this disk component is a regular ARDD.
(2) The vertex $v_1 \in \dot{D}_0^{(0)} \subseteq \dot{D}_1^{(0)} \subseteq \cdots \subseteq \dot{D}_n^{(0)}$ is such that the boundary paths pq and S_n from v_0 to v_1 satisfy the following: $\phi(p) \equiv a_1^{\epsilon_1} \cdots a_k^{\epsilon_k}$, $\phi(q) \equiv b_{k+1}^{\epsilon_{k+1}} \cdots b_{k+l}^{\epsilon_{k+l}}$ and $U[\phi(s_n)]^{-1}$ is a relator in $P_G = \langle S \mid \mathcal{R} \rangle$.

Let $\phi(s_n) \equiv x_1^{\eta_1} \cdots x_r^{\eta_r} \cdots x_{k+l}^{\eta_{k+l}}$, $\eta_j = \pm 1$, $x_j \in A \cup B$, and let $0 \leq r \leq k+l$ be the smallest index such that $x_r^2 \in \mathcal{R}_2$ and the edge e_r in s_n is not a boundary edge of a 2-cell. If no such r exists, we take $r = 0$. We attach a 2-cell to D_n along the edge e_r of s_n, such that $\phi(e_r) = x_r^{\epsilon_r}$. As a result we get a singular disk diagram D_{n+1} with boundary path pqs_{n+1}^{-1} such that s_{n+1} coincides with s_n everywhere except at the r-th edge, so $\phi(s_{n+1}) \equiv x_1^{\eta_1} \cdots x_r^{-\eta_r} \cdots x_{k+l}^{\eta_{k+l}}$. Note that this 2-cell is unique. If $r = 0$, let m, $(0 \leq m \leq k+l)$, be the smallest index such that the inverse of the subword $x_m^{\eta_m} x_{m+1}^{\eta_{m+1}}$ of $\phi(s_n)$ is in $\mathcal{L}(\mathcal{R}_{4*})$, i.e., $x_{m+1}^{-\eta_{m+1}} x_m^{-\eta_m} \in \mathcal{L}(\mathcal{R}_{4*})$. In this case we attach a 4-cell to D_n along the edges e_m and e_{m+1} of s_n, labeled with $x_m^{\eta_m}$ and $x_{m+1}^{\eta_{m+1}}$ respectively. Again, this 4-cell is unique. So, we've obtained a singular disk diagram $D_{n+1} = (D_{n+1}, \phi, pqs_{n+1}^{-1}, v_0)$ with boundary path pqs_{n+1}^{-1} such that the path s_{n+1} from v_0 to v_1 coincides with the path s_n everywhere except at the m-th and $m+1$-st edge. Not that in this case $\phi(s_n) < \phi(s_{n+1})$.

In both cases, ($r = 0$ or $r \neq 0$) the new complex D_{n+1} contains the old one D_n as a subcomplex. Finally, we define $\cup_{n \geq 0} D_n = \bar{D}_{[U]} = (\bar{D}_{[U]}, \phi, pqs_U^{-1}, v_0)$.

Intuitively, construction of $\bar{D}_{[U]}$ is simple: as we go along the path s_{n-1}, $(s_0 = pq)$ from v_0 to v_1 we examine labels of the edges of s_{n-1}. If the label is a symbol for which we have a relator in $\mathcal{R}_2$ we attach a 2-cell to the corresponding edge. After completing the attachments of all possible 2-cells which is done in an orderly fashion, we start attaching 4-cells, again from left to right. Using the notation of a (B, A)-pair from Proposition 2.1.2, we see that if D_{n+1} is obtained from D_n by attaching a 4-cell then $H(\phi(s_{n+1}^{-1})) \leq H(\phi(s_n^{-1}))$, therefore, this process will stop eventually. Note that we cannot go "below" the horizontal containing the vertex v_0 or "to the right" of the vertical containing the vertex v_1 (see Figure 2.8). So $D_{[U]}$ can have at most kl 4-cells. Finally, we obtain $D_{[U]}$ as a subdiagram of $\bar{D}_{[U]}$ by deleting some (if any) of the 2-cells on $\bar{s}_U$ so that $[\phi(s_U)]^{-1}$ is a C-geodesic.

We will prove (4) by induction on l. Let $l = 1$, that is, let $U \equiv a_1^{\epsilon_1} \cdots a_k^{\epsilon_k} b_{k+1}^{\epsilon_{k+1}}$ and $(M_1, \phi, p_1 q_1 s_1^{-1}, v_0)$ be a singular disk diagram for the relator $U[\phi(s_1)]^{-1}$, where $\phi(p_1) \equiv a_1^{\epsilon_1} \cdots a_k^{\epsilon_k}$, $\phi(q_1) \equiv b_{k+1}^{\epsilon_{k+1}}$ and let $W_1 \equiv \phi(s_1)$ represent the same element g as U. As noted earlier (Corollary 2.2.8) M_1 is regular ARDD so, we can place M_1 in the plane such that its A-edges lie vertically. Note that $\phi(s_1)$ must contain a B-symbol, otherwise M_1 will fail to be an ARDD. Let $\phi(s_1) \equiv A_s b_s S''$ where A_s is a nonempty A-word. If $|A_s| = m$, there is a unique 4-cell B_s in M_1 such that $b_s^{-1} a_{m+1}^{\epsilon_{m+1}} \in \mathcal{L}(\mathcal{R}_{4*})$. If any of the edges $e_{m+2}, \ldots, e_k, e_{k+1}$ has a cut vertex as its end vertex, s_1 fails to be geodesic. So the only possibility is that each of the end vertices of $e_2, \ldots, e_m$ is a cut vertex and $A_s \equiv a_1^{\iota_1} \cdots a_m^{\iota_m}$. If S' has a B-symbol then s_1 fails to be a geodesic. Thus S' is a nonempty A-word such that $|S'| = k - m$.

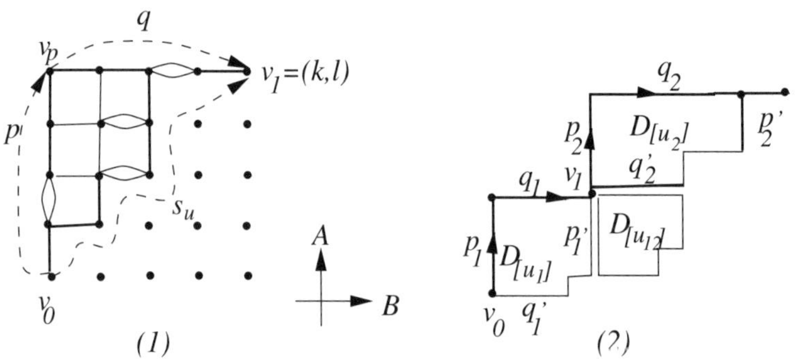

FIGURE 2.8

Suppose that (4) holds for all words $U' \equiv a_1^{\epsilon_1} \ldots a_k^{\epsilon_k} b_{k+1}^{\epsilon_{k+1}} \ldots b_{k+l}^{\epsilon_{k+l}}$ such that $l \leq n$. Let $U \equiv a_1^{\epsilon_1} \ldots a_k^{\epsilon_k} b_{k+1}^{\epsilon_{k+1}} \ldots b_{k+l+1}^{\epsilon_{k+l+1}}$ and W be such that $\bar{W} = \bar{U} = g \in G$. We can assume that U and W do not end with the same B-symbol (therefore the assumption is that W ends with an A-symbol). Let $b_w^{\epsilon_w}$ be the last B-symbol in W, that is, assume that $W \equiv W' b_w^{\epsilon_w} A_w$ or some nonempty A-word A_w. If A'_w is the C-geodesic obtained from A_w by changing some of the exponents of the A-symbols in A_w, following the base of the induction, there is a disk diagram $\bar{D}_{[A_w^{-1} b_w^{-1}]}$ which is a rectangular disk diagram such that labels on its B-chains are b_{k+l+1} and b_w. Let A''_w and A'_w be the labels of the A-chains of $\bar{D}_{[A_w^{-1} b_w^{-1}]}$. By the inductive hypothesis for the words U' and $W' A''_w$, there is a singular disk diagram $(M_n, \phi, pq_n s_n^{-1}, v_0)$ such that $\phi(pq_n) \equiv a_1^{\epsilon_1} \ldots a_k^{\epsilon_k} b_{k+1}^{\epsilon_{k+1}} \ldots b_{k+l}^{\epsilon_{k+l}}$, $\phi(s_n) \equiv W' A''_w$ that is a subdiagram of $\bar{D}_{[U']}$. Note that we can delete some 2-cells in the chain opposite to the A-chain in $\bar{D}_{[U']}$ labeled with a suffix of $a_1^{\epsilon_1} \cdots a_k^{\epsilon_k}$, such that we can find a sub-path of the boundary of $\bar{D}_{[U']}$ labeled with A''_w. It can be checked that identifying $\bar{D}_{[U']}$ and $\bar{D}_{[A_w^{-1} b_w^{-1}]}$ along this sub-path gives us $\bar{D}_{[U]}$. The singular disk diagram $D_{[U]}$ satisfies (4) by construction. □

Now, assume that $U \equiv A_1 B_1 A_2 B_2$ is a C-geodesic, where A_i's are some A-words and B_i's are some B-words. Let $U_i \equiv A_i B_i$, $(i = 1, 2)$. By the previous construction there will be two singular disk diagrams $(D_{[U_1]}, \phi, p_1 q_1 s_1^{-1}, v_0)$ and $(D_{[U_2]}, \phi, p_2 q_2 s_2^{-1}, v'_1)$. Let $D_{[U_1]} \vee D_{[U_2]}$ be a singular disk diagram obtained from $D_{[U_i]}$ by identifying the end vertex of q_1 (and s_1), say v_1 with the vertex v'_1. Following the construction in Theorem 2.2.2, let p'_1 be the maximal sub-path of s_1 such that $\phi(p'_1)$ is an A-word which is a suffix of $\phi(s_1)$. Similarly, let q'_2 be the maximal sub-path of s_2 such that $\phi(q'_2)$ is a B-word which is a prefix of $\phi(s_2)$. In order to simplify the argument, assume that $\phi(p'_1)$ and $\phi(q'_2)$ are C-geodesics, therefore, $U_{12} \equiv \phi(p'_1)\phi(q'_2)$ is a C-geodesic. Again, Theorem 2.2.2 gives a singular disk diagram $(D_{[U_{12}]}, \phi, p'_1 q'_2 s_{12}^{-1}, v_{12})$ where $\phi(p'_1)\phi(q'_2)\phi(s_{12})^{-1}$ is a relator over $P_G = \langle S \mid \mathcal{R} \rangle$, (see Figure 2.8 (2)). Identifying $D_{[U_{12}]}$ with $D_{[U_1]} \vee D_{[U_2]}$ along the sub-paths p'_1 and q'_2, one can see that the singular disk diagram obtained in this way has at most two disk components each with at least one 4-cell, such that these disk components are regular ARDD.

THEOREM 2.2.3. *If $U \equiv A_1B_1 \ldots A_nB_n$, $n \geq 1$, is a C-geodesic representing the element $g = \bar{U} \in G$, $G \in \mathcal{T}$, where A_i (B_j) are some nonempty A-words (B-words), then there is a unique singular disk diagram $\left(M_{[U]}, \phi, p_1q_1 \cdots p_nq_ns_U^{-1}, v_0\right)$ such that:*

1. *$M_{[U]}$ has at most n disk components each with at least one 4-cell, and these disk components are regular ARDD's.*
2. *There is a vertex v_1 in $\dot{M}_{[U]}^{(0)}$ such that for the boundary paths $p_1q_1 \cdots p_nq_n$ and s_n from v_0 to v_1, $\phi\left(p_1q_1 \cdots p_nq_n\right) \equiv U$ and $U[\phi(s_n)]^{-1}$ is a relator over P_G.*
3. *$\phi(s_n^{-1})$ is a C-geodesic representative of the element g^{-1} in G.*
4. *Every other geodesic representative of g will be a label of some path in $M_{[U]}$ from v_0 to v_1 with the length $|A_1| + |B_1| + \cdots + |A_n| + |B_n|$.*
5. *Every other singular disk diagram which includes two geodesics on its boundary representing the same element g in G, is a subdiagram of $\left(M_{[U]}, \phi, p_1q_1 \cdots p_nq_ns_U^{-1}, v_0\right)$.*

SKETCH OF PROOF. The proof is similar to that of Theorem 2.2.2 inducting on the number of subwords of the form A_kB_k. Note that in imitating the proofs of (4) and (5) of Theorem 2.2.2, the diagram $M_{[U]}$ has "extra" pieces as illustrated by $D_{[U_{12}]}$ in Figure 2.8 (2). These must be taken into account in imitating the discussion of the word W in that proof. □

NOTE 2.2.1. Since all our singular disk diagrams are reduced if v is a vertex in an ARDD, no two edges having v as initial vertex will have the same label. The same property holds even if v is an interior vertex in some disk diagram with d_4-degree equal to 6, or a boundary vertex with d_4-degree less than or equal to 3. (We will see that simple geodesic triangles in $\Gamma(G, A \cup B)$ have this property.) This allows us to see the 1-skeleton of ARDD's and the disk diagrams for simple geodesic triangles in $\Gamma(G, A \cup B)$ as part of the Cayley graph $\Gamma(G, A \cup B)$ identifying all vertices and edges in these diagrams with the corresponding vertices and edges in the Cayley graph. By saying that v is an *interior (boundary)* vertex in a geodesic triangle Δxyz we mean that v is an interior (boundary) vertex in a disk diagram $(D_\triangle, \phi, p_{xy}p_{yz}p_{zx}, v_0)$. This explains why these techniques of small cancellation theory can be applied when considering geodesic triangles in the Cayley graph. In general, this construction cannot be done.

PROPOSITION 2.2.2. *Let Δxyz be a simple geodesic triangle in $\Gamma(G, A \cup B)$, $G \in \mathcal{T}$, with geodesic sides p_{xy}, p_{yz} and p_{zx}. Let $(D_\triangle, \phi, p_{xy}p_{yz}p_{zx}, x)$ be a disk diagram over P_G for the relator $\phi\,(p_{xy})\phi(p_{yz})\phi(p_{zx})$. Then there is no interior vertex v in $D_\triangle$ such that $d_4(v) \geq 8$ and there is at most one interior vertex with d_4-degree equal to 6.*

PROOF. Using Lemma 2.2.1, the first sum in the left hand side of equation (2.1.2) can be decomposed and estimated as follows:

$$\dot{\sum_{v}}[2 - d_4(v)] = \sum_{v \in \{x,y,z\}} [2 - d_4(v)] + \sum_{v \in I}[2 - d_4(v)] \leq 3 + 3 = 6$$

where $I = Int(p_{xy}) \cup Int(p_{yz}) \cup Int(p_{zx})$. If there is an interior vertex v in $D_\triangle$ such that $d_4(v) \geq 8$, or more than one interior vertex in $D_\triangle$ with d_4-degree ≥ 6, we will have that

$$\sum_v^{\circ}[4 - d_4(v)] \leq -4$$

which implies that

$$\sum_v^{\cdot}[2 - d_4(v)] + \sum_v^{\circ}[4 - d_4(v)] \leq 2.$$

This contradicts (2.1.2). $\square$

PROPOSITION 2.2.3. *Let $\triangle xyz$ be a simple geodesic triangle in $\Gamma(G, A \cup B)$, $G \in \mathcal{T}$, with geodesic sides p_{xy}, p_{yz} and p_{zx}. Let $(D_\triangle, \phi, p_{xy}p_{yz}p_{zx}, x)$ be a disk diagram over P_G for the relator $\phi(p_{xy})\phi(p_{yz})\phi(p_{zx})$. Then*

(1) *there is no boundary vertex v in $D_\triangle$ such that $d_4(v) \geq 6$ and there is at most one boundary vertex with d_4-degree greater or equal to 4.*
(2) *$D_\triangle$ cannot contain an interior vertex of d_4-degree equal to 6 and a boundary vertex of d_4-degree greater than or equal to 4 at the same time.*

PROOF. It follows the same line of arguments as Proposition 2.2.2. $\square$

PROPOSITION 2.2.4. *Let $\triangle xyz$ be a simple C-geodesic triangle in $\Gamma(G, A \cup B)$, $G \in \mathcal{T}$ with C-geodesic sides p_{xy}, p_{yz} and p_{zx}. Let $(D_\triangle, \phi, p_{xy}p_{yz}p_{zx}, x)$ be a disk diagram over P_G for the relator $\phi(p_{xy})\phi(p_{yz})\phi(p_{zx})$. Then there is no boundary vertex v in $\dot{D}_\triangle$ such that $d_4(v) \geq 5$.*

PROOF. Let $v_5 \in \dot{D}_\triangle^{(0)}$ be such that $d_4(v_5) = 5$. Let p_1, p_2, p_3 be three paths in $D_\triangle$ with the initial vertex v_5 and terminal vertices v_1, v_2, v_3 respectively, $v_i \in \dot{D}_\triangle$, such that the labels $\phi(p_i)$ on each of these paths are non-empty A-words A_i (see Figure 2.9). We can assume that p_i has maximal length, i.e., $\phi(p_i)$ is not a prefix of $\phi(p_i')$for some path p_i' which begins at v_5.

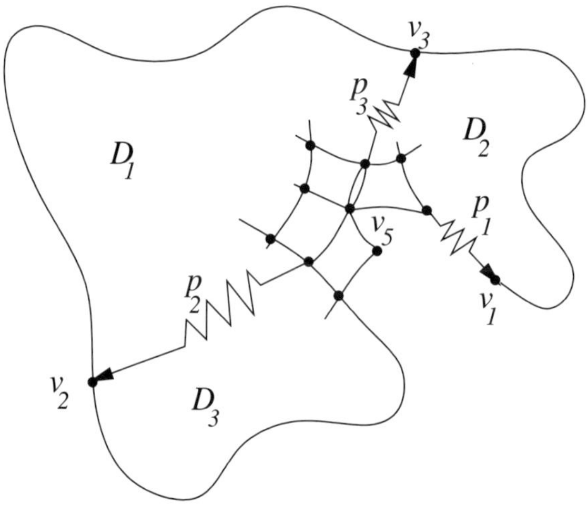

FIGURE 2.9

Note that exactly one of the p_i's begins with a boundary edge in $\dot{D}_{\triangle}^{(1)}$. We can assume that p_1 is such path. Let D_i be the subdiagram of $D_{\triangle}$ with the basepoint v_5 which does not contain the vertex v_i and it is one of the three (connected) components of $\overline{D_{\triangle} \setminus \{p_1 \cup p_2 \cup p_3\}}$. By Proposition 6.16 each D_i is an ARDD and no two of the vertices x, y, z can be in the same D_i. If D_1 has more than one A-chain disjoint from $p_3^{-1}p_2$ one of the sides of $\triangle$ will fail to be a C-geodesic. So, every D_i is a regular ARDD with x, y or z as one of its corners. We can take that $x \in \dot{D}_1$. Then $y \in \dot{D}_2$ and $z \in \dot{D}_3$ are corners of D_2 and D_3 respectively. It is clear that $z \neq 2$ and that z must be the second corner of the B-chain in D_3 which is incident to v_5. Therefore, x is a corner on the A-chain in D_1 opposite to the chain $p_3^{-1}p_2$ and y is the corner of the A-chain in D_2 opposite to $p_3^{-1}p_1$. But then, p_{yz} fails to be a C-geodesic at v_1. □

2.2.1. Classification of Simple C-geodesic Triangles. We now define *r-components* of certain disk diagrams $D_{\triangle}$. They are regular ARDD's, subdiagrams of $D_{\triangle}$ that can be one of the following three types:

a) Assume $(D_{\triangle}, \phi, p_{xy}p_{yz}p_{zx}, v_0)$ is a disk diagram for the relator $\phi(p_{xy})\phi(p_{yz})\phi(p_{zx})$ of a simple C-geodesic triangle $\triangle xyz$ such that there exists an interior vertex v_6 of $D_{\triangle}$ with $d_4(v_6) = 6$. Then there are three paths in the interior of $D_{\triangle}$, p_1, p_2 and p_3 with the initial vertex v_6 and the terminal vertices $v_i \in \dot{D}_{\triangle}$, $i = 1, 2, 3$, respectively such that $\phi(p_i) \equiv A_i$ where the A_i's are nonempty A-words. By Propositions 6.16 and 6.17, the closure of any one of the three components of $D_{\triangle} \setminus \{p_1 \cup p_2 \cup p_3\}$ is a regular ARDD which is a subdiagram of $D_{\triangle}$. We call this subdiagram a *regular ARDD component of* $D_{\triangle}$ or simply an *r-component* of $D_{\triangle}$.
b) Similarly, if v_4 is a boundary vertex of $D_{\triangle}$ with $d_4(v_4) = 4$, there are at most three paths p'_j, $j = 1, 2, 3$ of $D_{\triangle}$ with v_4 as the initial vertex and the terminal vertices $v'_j \in \dot{D}_{\triangle}$, $j = 1, 2, 3$ respectively. In this case $\phi(p'_j) \equiv A'_j$, where A'_j are some nonempty A-words. Then the closure of $D_{\triangle} \setminus \{p'_1 \cup p'_2 \cup p'_3\}$ has two regular ARDD components which are subdiagrams of $D_{\triangle}$. These are r-components.
c) If in (b) there are only two paths p'_1 and p'_2 (i.e., the third path is empty) then the closure of $D_{\triangle} \setminus \{p'_1 \cup p'_2\}$ has three regular ARDD components which are subdiagrams of $D_{\triangle}$. These are r-components.

So, the shape of a simple C-geodesic triangle $\triangle xyz$ in $\Gamma(G, A \cup B)$ is determined by the number of r-components of $D_{\triangle}$. With this number and the number of chains on the boundary of $D_{\triangle}$ (which is at most 6) we can classify the simple C-geodesic triangles in $\Gamma(G, A \cup B)$, $G \in \mathcal{T}$ as follows:

(i) If the number of r-components of $D_{\triangle}$ is 0 then $D_{\triangle}$ does not contain a 4-cell and consequently two of x, y, z will coincide. In this case $D_{\triangle}$ is simply a 2-cell.

If the number of r-components is 1, then $D_{\triangle}$ does not contain interior vertex of d_4-degree equal to 6 or boundary vertex of d_4-degree equal to 4.

If the number of chains on the boundary $D_{\triangle}$ is 5 no two of the vertices x, y, z can coincide (see Figure 2.10 (1)).

(ii) If the number of r-components is greater than 1, $D_{\triangle}$ can contain an interior vertex of d_4-degree equal to 6 in which case the number of r-components is 3, or a boundary vertex of d_4-degree equal to 4 in which case it has at most two r-components (see Figure 2.10 (2) and (3))

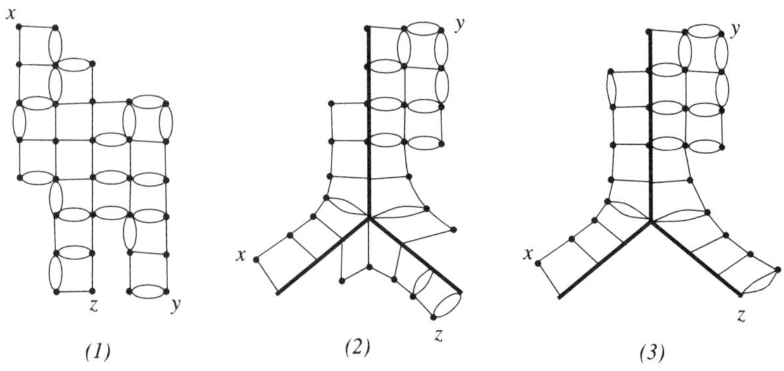

FIGURE 2.10

We define the of a rectangle to be the maximum of the lengths of its sides and the size of a simple geodesic triangle $\triangle$ in $\Gamma(G, A \cup B)$, $G \in \mathcal{T}$ to be $\max\{\text{distance}(x, A) \mid A$ is the union of two sides of $\triangle$ and $x \in \triangle\}$. Then it is clear that the size of a geodesic triangle in $\Gamma(G, A \cup B)$, $G \in \mathcal{T}$ no more than twice the maximum of the sizes of its r-components.

DEFINITION 2.2.2. Let (M, ϕ, p, v_0) be a singular disk diagram over P_G, $G \in \mathcal{T}$ and let C_M be the set of all cut vertices of M. We say that $v \in C_M$ is a *simple cut vertex* if $M \setminus \{v\}$ has two connected components. Otherwise v is a *multiple cut vertex.*

We also define a *disk component graph* Γ_M of M as follows. The set of vertices of Γ_M, $(V(\Gamma_M)$ consists of the set of connected components of M together with a set of all multiple cut vertices of M. There will be an (unoriented) edge e in the set of edges $E(\Gamma_M)$ of Γ_M having δ_1 and δ_2 as its end vertices iff

(i) either δ_1 and δ_2 are connected components of M having a simple cut vertex in common, or
(ii) δ_1 is a multiple cut vertex and δ_2 is a connected component with δ_1 on its boundary (see Figure 2.11).

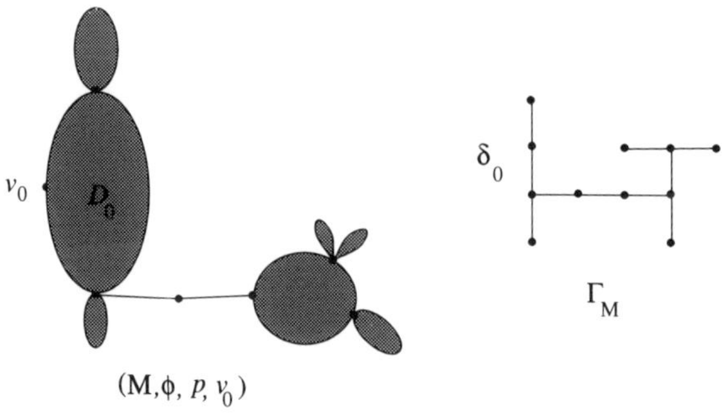

FIGURE 2.11. The figure on the right is the disk component graph Γ_M of the singular disk diagram (M, ϕ, p, v_0) shown on the left.

It follows from the definition that Γ_M is a tree and it will consist of a single vertex iff (M, ϕ, p, v_0) is a disk diagram. Let D_0 be the disk component of (M, ϕ, p, v_0)

to which v_0 belongs and let δ_0 be the vertex in Γ_M representing D_0. The boundary path p in M gives a rise to a path π in Γ_M with initial and terminal vertex δ_0, such that π passes through all vertices of Γ_M. Note that if $\delta \neq \delta_0$ is a vertex in Γ_M with degree 1 and e is the edge in $E(\Gamma_M)$ incident to δ, then π (as a word in the group alphabet $E(\Gamma_M)$) will have a subword ee^{-1} and δ must represent a disk component of (M, ϕ, p, v_0).

As mentioned earlier (Theorem 1.4.1), it is important to know when a group G in our class $\mathcal{T}$ has a free abelian subgroup of rank 2. The following theorem is the main step in the connection between the existence of such subgroups and solvability of the so called "tiling problem" (see 3.1).

THEOREM 2.2.4. *Let (M, ϕ, p_0, v_0) be a singular disk diagram for a freely reduced relator $UVU^{-1}V^{-1}$, where U and V are C-geodesics in $\Gamma(G, A \cup B)$ with no common prefix or suffix. If $\bar{U}$ and $\bar{V}$ are not powers of the same element in G then (M, ϕ, p_0, v_0) is a rectangular disk diagram.*

PROOF. Let (M, ϕ, p_0, v_0) be a singular disk diagram for the relator $UVU^{-1}V^{-1}$ over P_G. There will be four paths on the boundary of M, p, q, p', q', such that $p_0 = pq'(p')^{-1}q^{-1}$, $\phi(p) \equiv \phi(p') \equiv U$ and $\phi(q) \equiv \phi(q') \equiv V$. Let v_p be the end vertex of p and the initial vertex of q', v_q the end vertex of q and the initial vertex of p' and v' the end vertex of q' and p'. We will have four corresponding paths in Γ_M, $\pi_p, \pi_q, \pi_{p'}$ and $\pi_{q'}$ respectively. Of course, if some of the paths $p, q, \ldots$ is included in the boundary of a disk component of M, the corresponding paths $\pi_p, \pi_q, \ldots$ will be reduced to vertices in Γ_M.

Assume that $\pi_p = e_1 \ldots e_k e_k^{-1} \ldots e_n$, $n \geq 0$, contains a sub-path $e_k e_k^{-1}$ and let δ_k be the vertex in Γ_M which is incident only to e_k.

Then δ_k will represent a disk component D_k in (M, ϕ, p_0, v_0) and p will fail to be a geodesic. So each of the paths $\pi_p, \pi_q, \pi_{p'}, \pi_{q'}$ is either reduced to a vertex in Γ_M or it is represented by a freely reduced word in the group alphabet $E(\Gamma_M)$. With this remark in mind, we will break up the proof of this theorem into several lemmas. □

Let $\delta_0, \delta_p, \delta_q$ and δ' be vertices in Γ_M representing the disk components (or multiple cut vertices) to which v_0, v_p, v_q and v' belong, respectively.

LEMMA 2.2.2. *If all of $\delta_0, \delta_p, \delta_q$ and δ' coincide, then (M, ϕ, p_0, v_0) is a rectangular disk diagram.*

PROOF OF LEMMA 2.2.2. First we will show that for every interior vertex v in M, $d_4(v) = 4$.

Assume that there is an interior vertex v_6 in $M^{(0)}$ such that $d_4(v_6) \geq 6$. There are three paths p_1, p_2 and p_3 with an initial vertex v_6 and terminal vertices v_1, v_2 and v_3 on the boundary of M such that the labels of these paths are some non-empty A-words (see Figure 2.11 (1)). We can assume that each of these paths is maximal in the sense that it is not a sub-path of a path with the same properties.

Let D_i be the subdiagram of M which does not contain the vertex v_i ($i = 1, 2, 3$), on its boundary and it is one of the three (connected) components of $M \setminus \{p_1 \cup p_2 \cup p_3\}$ (see Figure 2.11 (a)). We can choose v_6 such that D_3 does not contain an interior vertex with d_4-degree greater than four and such that $v_0 \in \dot{D}_3$. Note that each of the D_i's contains at least two 4-cells and at least one A-chain which does not contain the vertex v_6. If v' is in D_3, either U or V will contain some

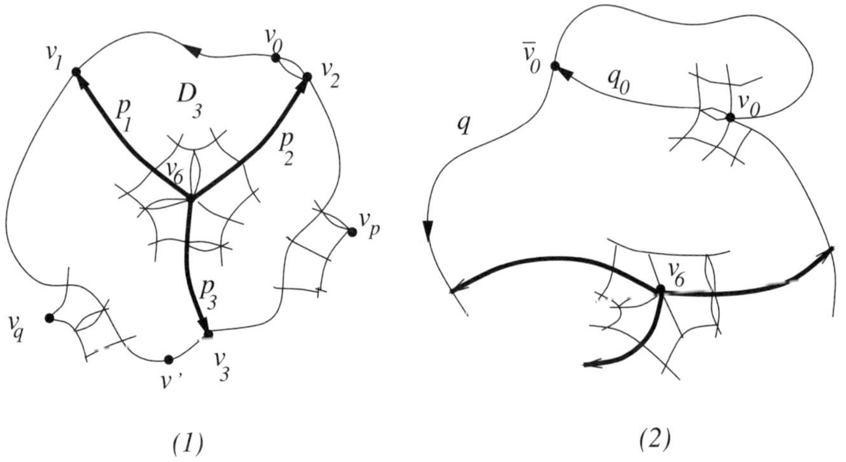

FIGURE 2.12

of these chains and fail to be a geodesic. Therefore v' must be on the boundary of D_2 or D_1. Let $v' \in \dot{D}_2$. If D_1 contains an interior vertex v_6' with $d_4(v_6') \geq 6$, again there will be three paths p_1', p_2' and p_3' with an initial vertex v_6' and terminal vertices v_1', v_2' and v_3' ($v_i' \neq v_j$, $\forall i, j = 1, 2, 3$) on the boundary of D_1. We can choose v_6' such that one of the closed components of $\overline{D_1 \setminus \{p_1' \cup p_2' \cup p_3'\}}$ does not contain an interior vertex with d_4-degree greater than four. But then D_1 will contain at least two A-chains so one of U and V will fail to be a C-geodesic. Similarly we can show that there is no interior vertex in D_2 with d_4-degree greater than 4. We get to the same conclusion if there exist a vertex on p_i, (as a path included in the boundary of D_i, with d_4-degree greater than two.

Having an A-chain in D_1 (D_2) which has no common vertex with $p_2 \cup p_3$ ($p_1 \cup p_3$) implies that one of the endpoints of this chain must be the vertex v_p (v_q) and that both U and V have A-words as their prefixes (see Figure 2.12 (1)).

Now we have two choices for $d_4(v_0)$:

(1) $d_4(v_0) \leq 2$
(2) $d_4(v_0) \geq 4$

In the second case there will be a path q_0 in D_3 having v_0 as its initial vertex and $\bar{v}_0 \in \dot{D}_3$ as its terminal vertex, such that $\phi(q_0)$ is some A-word (see Figure 2.12 (2)). But $v_q \notin D_3$ therefore, V fails to be a geodesic. Similarly we can show that there is no boundary vertex in M with d_4-degree greater than 3. As a consequence of this, $d_4(v_0) \leq 2$ and each of the subdiagrams D_i is a regular ARDD. Note that if one of the D_i's is not a disk diagram then it contains more than one A-chain on its boundary.

Having D_3 as a regular ARDD with one A-chain included in the boundary path $p_1^{-1}p_2$, let n_3 be the distance between $p_1^{-1}p_2$ and the unique A-chain in D_3 "opposite" to $p_1^{-1}p_2$. We can think of n_3 as a measurement of the "height" of D_3. Similarly we can introduce n_1 and n_2 as heights of D_1 and D_2. Because each of the D_i's has at least one 4-cell, each of the numbers n_i is positive (see Figure 2.13).

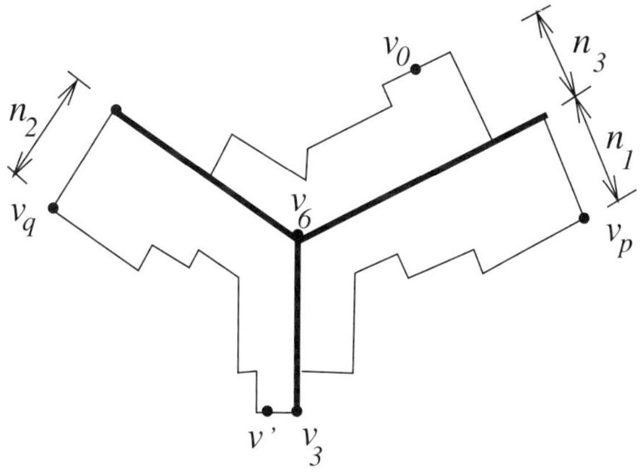

FIGURE 2.13

Let q'' be the sub-path of q' having v_3 as its initial vertex and v' as its terminal vertex and n_2'' be the number of B-symbols in the word $\phi(q'')$. We denote with $|X|_B$ the number of B-symbols in the word X, counted with multiplicities.

We have the following relations between the n_i's:

$$(*) \qquad \begin{aligned} n_3 + n_1 &= |U|_B \\ n_3 + n_2 &= n_1 + n_2'' = |V|_B \end{aligned}$$

Having that p' is entirely in D_2, we will get:

$$(**) \qquad n_2 - n_2'' = n_3 + n_1$$

But then $(*)$ and $(**)$ will give $n_2 - n_3 = n_3 + n_2$ or $n_3 = 0$ which is a contradiction. Thus, we conclude that there is no interior vertex in M with d_4-degree greater than four. Following the same line of arguments, we can show that if v is a boundary vertex of M then $d_4(v) \leq 3$ and conclude that M is an ARDD.

Assume that $d_4(v_0) = 2$, i.e., both U and V have an A-word as a prefix. There are at least two B-chains in M (separated by at least one A-chain in M), so each of the vertices v_p and v_q must be an end vertex for a B-chain in M, i.e., U and V both end with a B-word as a suffix. If v_0 does not belong to an A-chain, one of p or q will fail to be a geodesic. So v_0 is an interior vertex of an A-chain and v_p and v_q have d_4-degree equal to 1. Both v_p and v_q must belong to A-chains too, otherwise either q' or p' will fail to be a C-geodesic (see Figure 2.14).

Let t_0 be the shortest path in M with initial vertex v_0 and terminal vertex $v'' \in \dot{M}$, such that $\phi(t_0)$ is a non-empty B-word. Assume that $v'' \in q'$. Note that $v'' \neq v'$. Let q_1' (q_2') be the sub-path of q' with initial vertex v_p (v'') and terminal vertex v'' (v'). Then

$$|\phi(p)|_B = |t_0| + |\phi(q_1')|_B = |\phi(p')|_B = |\phi(q)|_B - |t_0| - |\phi(q_2')|_B,$$

i.e., $|t_0| = |\phi(t_0)|_B = 0$ which is a contradiction. So, $d_4(v_0) = 1$ and U and V have prefixes from different sets of symbols. With a similar argument we can exclude the possibility of a boundary vertex in M, having d_4-degree equal to 3. Thus, either U is a B-word and V is an A-word or vice versa. □

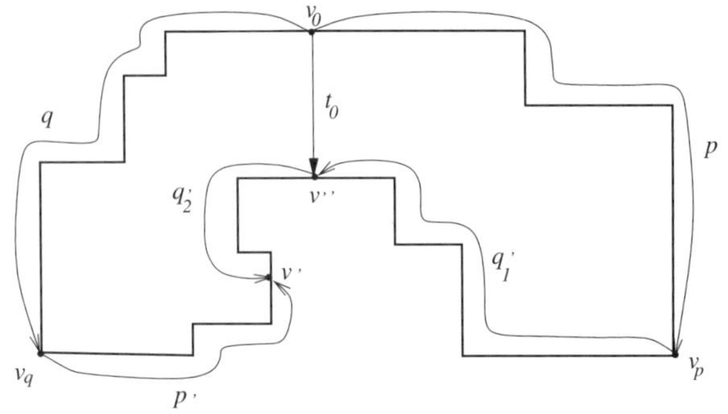

FIGURE 2.14

LEMMA 2.2.3. *If all of $\delta_0, \delta_p, \delta_q$ and δ' are distinct, then U and V represent powers of the same element in G.*

PROOF OF LEMMA 2.2.3. We will represent each of the paths $\pi_p, \pi_q, \pi_{p'}, \pi_{q'}$ either by a single edge or by at most three edges connecting some of the vertices $\delta_0, \delta_p, \delta_q$ or δ'. For example, if π_p (π_q) does not contain δ' and δ_q (δ_p), Γ_M can be represented as in Figure 2.15.

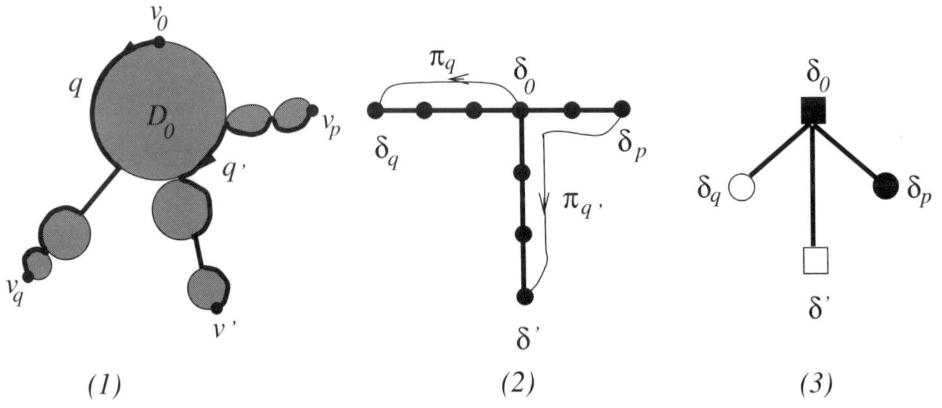

FIGURE 2.15. The figure (1) represents a singular disk diagram (M, ϕ, p, v_0). The orientation of the edges and labels of the paths p and p' are not indicated; (2) represents the disk component graph Γ_M and (3) shows how Γ_M is represented in Fig. 2.15.

In this case, the boundary path p has no edges in common with the disk component D'. All other vertices in π_p which may represent disk components (or multiple cut vertices) in (M, ϕ, p_0, v_0) are not indicated. Assuming that δ_q is the "left vertex", δ_0 the "top vertex", δ' the "bottom vertex" and δ_p the "right vertex" in every four vertices below, and that π_p (π_q) is the minimal path having δ_0 and δ_p (δ_q) as its end vertices, while $\pi_{p'}$ ($\pi_{q'}$) is the minimal path having δ_q (δ_p) and δ' as its end vertices, the possibilities for Γ_M are presented in Figure 2.16.

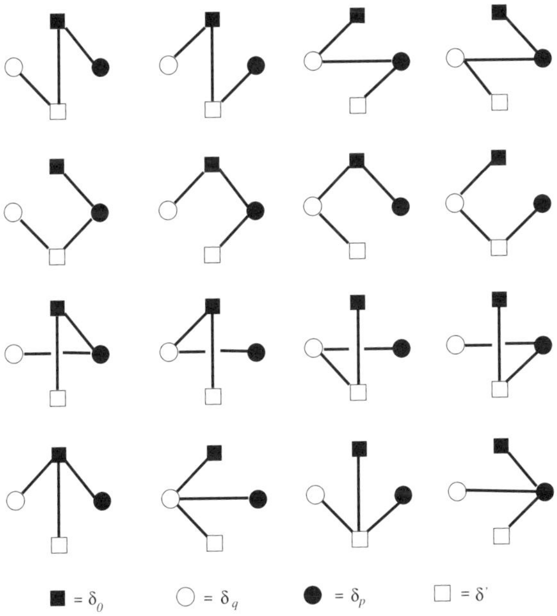

FIGURE 2.16

Assume $\deg(\delta_0) = 1$ and let δ_0 represent a disk component in (M, ϕ, p_0, v_0). Having that a prefix of a C-geodesic is a C-geodesic, there is a vertex v_1 on the boundary of D_0 such that spelling the symbols of the two boundary paths from v_0 to v_1 we get words which are prefixes of U and V. By Corollary 2.2.4, U and V have the same prefix which contradicts the assumption. A similar conclusion will follow when $\deg(\delta') = 1$. So the only non-trivial possibilities are the first two cases in the first row of Figure 2.16. We will discus the first case. The second one is "symmetrical" and a similar argument will hold.

Note that if δ_p or δ_q represent disk components with only one 2-cell, then there are subwords U' and V' of U and V respectively, such that $U \equiv U'x$ and $V \equiv xV'$, where $x^2 \in \mathcal{R}_2$. In this case the relation $UV = VU$ can be reduced to $U'V' = V'U'$. So, if D_p and D_q are disk components in M represented by δ_p and δ_q respectively, we can assume that they contain at least one 4-cell. By Corollary 2.2.8 we have that D_p and D_q are regular ARDD's and consequently U and V begin with an A-symbol and end with a B-symbol.

Before considering different cases, note that using the same argument from Corollary 2.2.8, every disk component (subdiagram) $D = (D, \phi, s_p s_q, v)$ of M, such that $v \in M^{(0)}$ is a cut vertex in M, s_p is a sub-path of p (or p'), with initial vertex v and s_q is a sub-path of q (or q') with terminal vertex v, is a regular ARDD. Therefore (following Proposition 2.2.2), we can place these disk components of M in the plane such that each edge has unit length, their A-edges lie horizontally and their B-edges vertically. If t is a boundary path included in M that connects two disk components of M, we will place t in the plane such that its initial and its terminal vertex are opposite corners of a rectangle with lengths of its sides $|\phi(t)|_A$ and $|\phi(t)|_B$.

Now we have three subcases.

Case 1. v_0 and v' are cut vertices in the components represented by δ_0 and δ' respectively.

As we noted above, we place the disk components of M in the plane such that their A-edges lie horizontally and their B-edges vertically. Let p_0^1 denote the path in (M, ϕ, p_0, v_0) connecting v_0 and v' and let π_0^1 be the corresponding path in Γ_M connecting δ_0 and δ'. By Corollary 2.2.4 every vertex in π_0^1 can represent only a cut vertex in a component which does not contain cells.

Let q_1' be the sub-path of q' connecting v_p and v_0 and let q_1 be the sub-path of q connecting v' and v_q. Then $V \equiv \phi(q_1')\phi(p_0^1) \equiv \phi(p_0^1)\phi(q_1)$. If M' (M'') is a subdiagram of M having p (p') and q_1' (q_1) on its boundaries, by Theorem 2.2.3 (M', ϕ, pq_1', v_0) and $(M'', \phi, p'q_1, v_q)$ coincide.

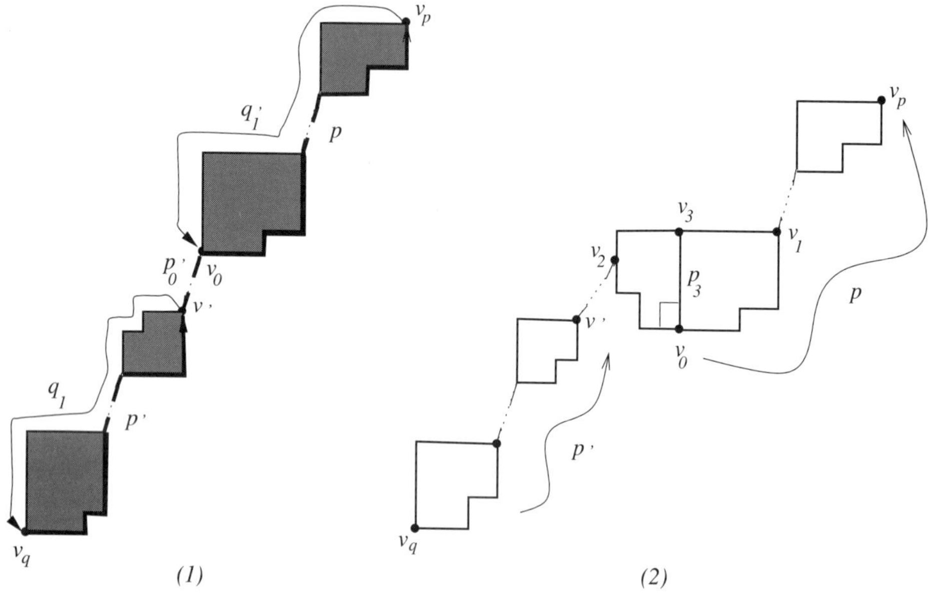

FIGURE 2.17. We illustrate the cases 1 and 2 with (1) and (2) respectively. Only few distinguished vertices are indicated and some of these are labeled.

Note that $\phi(p_0^1)$, like V, must begin with an A-symbol and end with a B-symbol. Therefore it contains a subword of the form AB where A is some nonempty A-word and B is some nonempty B-word. Assume that $\phi(p_0^1)$ is of the form AB for some nonempty A-word A and some nonempty B-word B. Then $\phi(q_1')$ has $\phi(p_0^1)$ as its prefix and $\phi(q_1)$ ($\equiv \phi(q_1')$) has $\phi(p_0^1)$ as its suffix. Hence V, being a C-geodesic, coincides with $\phi(p_0^1)V'\phi(p_0^1)\phi(p_0^1)$ when we read the label of q' and with $\phi(p_0^1)\phi(p_0^1)V'\phi(p_0^1)$ when we read the label of q (V' is a subword of V). We see that either V' is the empty word or we can find a subword V'' of V' such that $V' \equiv \phi(p_0^1)V''\phi(p_0^1)$. This process certainly ends (V has finite length), so V is a power of $V_0 = \phi(p_0^1)$, that is, there exists $l \geq 1$ such that $V = V_0^l$. The same conclusion can be reached if $\phi(p_0^1)$ is a power, i.e., if $\phi(p_0^1) \equiv (A_1B_1 \cdots A_kB_k)^n$, for $n \geq 2$.

Now let v_1 be the vertex on q_1' such that the sub-path of q_1' beginning with v_p and ending with v_1 has the label $\phi(p_0^1)$. Again, by Theorem 2.2.3, v_1 is necessarily a cut vertex. If we label with p_1 the sub-path of p with initial vertex v_1 and terminal vertex v_p, we see that $\phi(p_0^1)\phi(p_1)$ is a relator over $P_G = \langle S \mid \mathcal{R} \rangle$. By Theorem 2.2.3, we conclude that $V = V_0^l$, $U = [\phi(p_1)]^l$ and that $\bar{V}_0\phi(p_1) = 1$, that is, $\bar{U}$ and $\bar{V}$ are powers of the same element $\bar{V}_0$.

Case 2. v_0 is not a cut vertex in a disk component represented by δ_0 and v' is a cut vertex in a disk component represented by a disk component δ'.

Note that if D_0 has an interior vertex v with $d_4(v) > 4$, then there are at least three A-chains on the boundary of D_0. Having that the boundary of D_0 consists of three paths such that two of them are sub-paths of p and q with the same initial vertex v_0 and the third one is a sub-path of q', we see that one of the A-chains in D_0 is not a sub-path of any of these sub-paths. Similar conclusion (contradiction) can be reached if we assume that there exists a boundary vertex v with $d_4(v) = 4$. Hence, the disk component D_0 must be an ARDD.

Since U and V begin with an A-symbol, we have that $d_4(v_0) \leq 2$. Assume that $d_4(v_0) = 2$. The boundary of D_0 consists of three C-geodesic paths: a sub-path of p with initial vertex v_0 and terminal cut vertex v_1, a sub-path of q' with initial vertex v_1 and terminal cut vertex v_2 and a sub-path of q with initial vertex v_0 and terminal vertex v_2 (see Figure 2.17 (2)).

There is a unique C-geodesic path t_3 in D_0, having v_0 as its terminal vertex and some $v_3 \in \dot{D}_0$ as its initial vertex, such that $\phi(t_3)$ is a non-empty B-word. Here we label with q_p' the sub-path of q' having v_p as its initial vertex and v_3 as its terminal vertex. Let q_1 denote the same path as in Case 1. Following Theorem 2.2.3, subdiagrams $(M', \phi, pq_p't_3, v_0)$ and $(M'', \phi, p'q_1, v_q)$ will coincide. Note that $\phi(q_p')$, $\phi(q)$ and $\phi(q_1)$ all begin with the same A-symbol. Also $\phi(t_3)$, $\phi(q')$ and $\phi(q)$ all end with the same B-symbol. Now, the last edge of t_3 and the first edge of q are labeled with B-symbol and A-symbol respectively, and there is a 4-cell in D_0 with these two edges on its boundary. Because the labels of the edges of q incident to v' are the same as the labels of these two edges incident to v_0, q fails (at v') to be a C-geodesic. So, $d_4(v_0) = 1$ and Case 2 reduces to Case 1.

Case 3. v_0 and v' are not cut vertices in the disk components represented by δ_0 and δ'.

Following the same argument as in Case 2, we see that both D_0 and D' must be ARDD's. We need some notation. The vertices v_0, v_1, v_2 and v_3 on the boundary of D_0 and the paths t_3 and q_p' are same as in Case 2. Let v_1' be one of the two cut vertices in $\dot{D}'$ where q and q' are "entering" D' (see Figure 2.18 (1)). Let p_2^1 be a sub-path of q (or q') with initial vertex v_2 and terminal vertex v_1'. There is a unique C-geodesic path t' in D' such that v' is its initial vertex, some $v_2' \in \dot{D}'$ is its terminal vertex and $\phi(t')$ is a non-empty A-word. Let v^* be the unique vertex of t' such that the C-geodesic path t_1' having v_1' as its initial vertex and v^* as its terminal vertex, has a label some B-word. We label with q_2 the sub-path of q having v_2' as its initial and v_q as its terminal vertex (see Figure 2.18 (1)). Finally, on t_3 there will be a unique vertex v_* such that the C-geodesic path t_2, having v_2 as its initial and v_* as its terminal vertex, has label some A-word (see Figure 2.18 (1)).

Again, by Theorem 2.2.3, subdiagrams $(M', \phi, pq_p't_3, v_0)$ and $(M'', \phi, p't'q_2, v_q)$ coincide. If we compare B-lengths of the words $\phi(q')$ and $\phi(q)$ we see that

$$|\phi(q')|_B = |\phi(q)|_B - |\phi(t_*)|_B$$

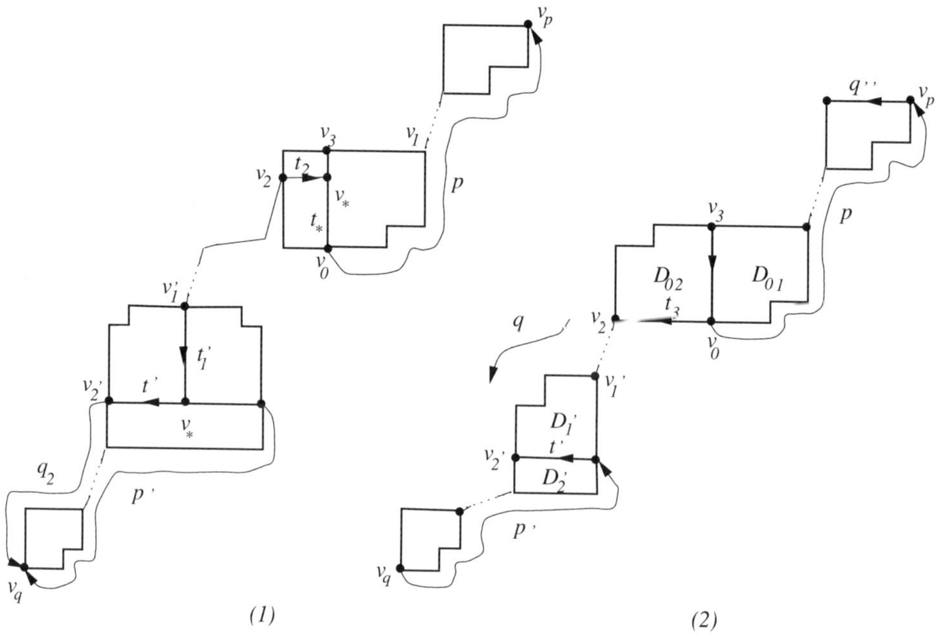

FIGURE 2.18

where t_* is the unique C-geodesic path connecting v_* and v_0, (therefore a sub-path of t_3). Thus $|\phi(t_*)|_B = 0$, i.e. $v_* = v_0$ and $d_4(v_2) = 1$. Similarly, by comparing the A-lengths of q and q' we will get that $v^* = v'$ (see Figure 2.18 (2)). The path t_3 splits the disk component D_0 into two subdiagrams D_{01} and D_{02}. We take $v_1 \in \dot{D}_{01}$.

Similarly, the path t' splits the disk D' into two subdiagrams D'_1 and D'_2. We take that $v'_1 \in \dot{D}'_1$. Because $\phi(q'_p)\phi(t_3)$ and $\phi(t'\phi(q_2)$ are C-geodesics, by Theorem 2.2.3, D_{01} coincides with D_q and D'_2 with D_p. Obviously, $\phi(t')$, $\phi(q_0)$ and $\phi(q')$ ass start with the same A-symbol. Nevertheless, if $\phi(q_0)$ is a proper prefix of $\phi(t')$ then q' will fail to be C-geodesic at v_2. This follows from the fact that D'_1 with its boundary made of C-geodesics $(t')^{-1}t'_1$ and the sub-path of q with v'_1 and v'_2 as its initial and terminal vertices is uniquely determined. The same conclusion can be reached if $\phi(t'_1)$ is a proper prefix of $\phi(t_3)$ or vice versa. So, D'_1 coincides with D_{02}.

Let q_0 be the sub-path of q with initial vertex v_0 and terminal vertex v_2 (therefore $\phi(q_0)$ is some A-word). There will be a sub-path q'' of q_p, with initial vertex v_p and terminal vertex some $v'' \in \dot{D}_p$ such that $\phi(q'') \equiv \phi(q_0) \equiv \phi(t')$. Let q'_1 be the sub-path of q_p with initial vertex v'' and terminal vertex v_3 (see Figure 2.18 (2)).

If $\phi(q_0) \equiv A_1(\equiv \phi(q''))$, $\phi(t_3) \equiv B_1$, $\phi(q'_1) \equiv V_1$ and $\phi(p^1_2) \equiv V_3$ then

$$\phi(q') \equiv A_1V_1V_2V_3B_1 \equiv \phi(q) \equiv A_1V_3V_2V_1B_1$$

where $V_2 \equiv \phi$(the sub-path of q' with initial vertex v_3 and terminal vertex v_2). Inducting on the sum of the lengths of the V_i, there will be a C-geodesic word V_0 such that $V_i \equiv V_0^{n_i}$ for some $n_i \in \mathbb{Z}$. But then $\overline{UA_1V_0^{n_1}B_1} = 1$ and $V \equiv A_1V_0^{n_1+n_2+n_3}B_1$. The label on the boundary of D_{02} reads $V_0^{n_2}A_1^{-1}B_1^{-1}$ so $\bar{U} =$

$\overline{B_1^{-1}V_0^{-(n_1+n_2)}B_1}$ and $\bar{V} = \overline{B_1^{-1}V_0^{n_1+2n_2+n_3}B_1}$, i.e., U and V represent powers of the same element in G. □

LEMMA 2.2.4. *If $\delta_0 \neq \delta'$ and exactly one of δ_p and δ_q coincide with either δ_0 or δ' then U and V are powers of the same element in G.*

PROOF OF LEMMA 2.2.4. We have fourteen possibilities for Γ_M presented in Figure 2.19. (Notation is the same as in Figure 2.16.)

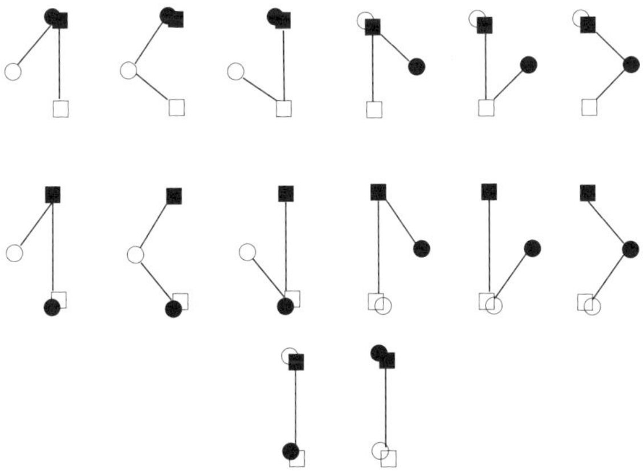

FIGURE 2.19

Note that whenever δ_0, (δ') has not been identified with δ_p or δ_q and has $deg_\Gamma = 1$, U and V have a common prefix (suffix). So we can exclude these eight cases. Third case in Figure 2.19 follows from the third case of the previous lemma taking $v_1 = v_p$. Using symmetry, we are left with the last two cases in Figure 2.19. But these two cases are covered by the Case 3 in previous lemma if we assume that D_{01} coincides with D_p and D_2' with D_q. □

LEMMA 2.2.5. *The possibility of $\delta_0 = \delta'$ and at least one of δ_p or δ_q is different from δ_0 cannot occur.*

PROOF OF LEMMA 2.2.5. We have the following possibilities for Γ_M (see Figure 2.20).

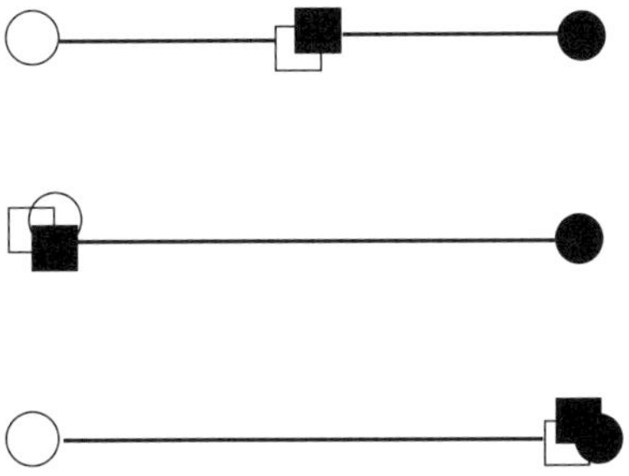

FIGURE 2.20

Case 1. Let $D_0 = D'$ be the disk component represented by $\delta_0 = \delta'$ in Γ_M. There are three subcases to be considered. If both v_0 and v' are cut vertices in D_0 (which is necessarily an ARDD) they cannot be the end vertices of the same chain. But then this case, reduces to Case 1 of Lemma 2.2.3 where we take $|p_0^1| = 0$. Now, the case of v_0 being a cut vertex but v' not being a cut vertex, leads to a contradiction because $\phi(p')$ must have a B-word as its suffix. Finally, if none of v_0 and v' are cut vertices, since $\phi(p)$ and $\phi(q)$ must begin with an A-symbol and end with a B-symbol, $d_4(v_0) = d_4(v') = 2$. But this implies that there is a boundary path in D_0 which is a sub-path of p' (or q') which fails to be a C-geodesic.

Case 2. Assume that v_1 is a cut vertex in D_0 where p "exits" D_0. Then the boundary of D_0 contains four C-geodesic paths: $\bar{p}$ with initial vertex v_0 and terminal vertex v_1 (a sub-path of p), q, p' and $\bar{q}'$ (a sub-path of q' with initial vertex v_1 and terminal vertex v'). It is not difficult to see that the disk component D_0 $(= D' = D_q)$ must be an ARDD (not necessarily a regular one). There are at least two B-chains and at least two A-chains on the boundary of D_0. Let χ_q be the B-chain which belongs to q, χ_q' a B-chain opposite to χ_q, $\chi_{p'}$ an A-chain which belongs to p' and χ_0 the A-chain with v_0 as one of its vertices. Note that $d_4(v_q)$ must equal 1 (otherwise one of q or p' will fail to be a C-geodesic). So, v_q is a common vertex of χ_q and $\chi_{p'}$ which implies that U begins with an A-symbol and V terminates with a B-symbol. Consequently $d_4(v_0) = 2$ and $v' \in \chi_q'$. Look at the sub-path t^* in D_0 with initial vertex v' and terminal vertex v^* which belongs to q, such that $\phi(t^*)$ is some nonempty A-word. If q^* is a sub-path of q with initial vertex v^* and terminal vertex v_q, the subdiagram $(M_q, \phi, p't^*q^*, v_q)$ is uniquely determined. Hence, there must be a boundary vertex v_1^* in q^* (and p') that "corresponds" to the boundary vertex v_1 in p. But then, $\delta_q \neq \delta_0$ contrary to the assumption that we started with (see Figure 2.21).

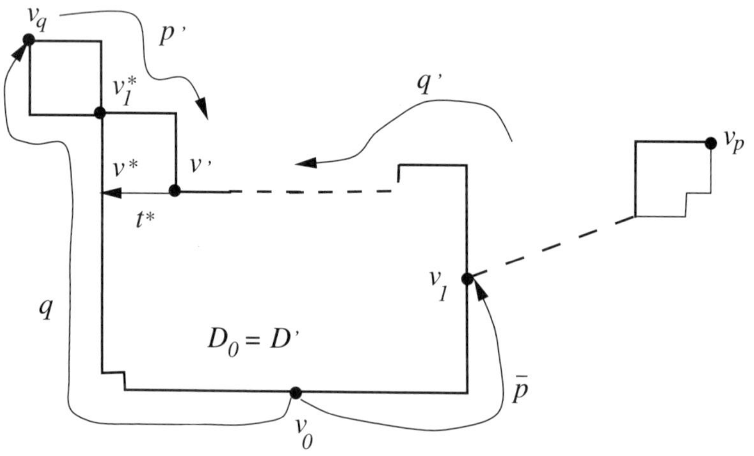

FIGURE 2.21

Case 3. The proof of this case follows the same line of arguments as Case 2. □

With this we exhausted all possibilities for δ_0, δ_p, δ_q and δ' which concludes the proof of Theorem 2.2.4.

CHAPTER 3

Tilings

3.1. Tilings

To every ARDD (D, ϕ, p, v_0) over P_G, $G \in \mathcal{T}$, and a boundary vertex $v \in \dot{D}^{(0)}$ we will associate a RDD (D_v, ϕ, p_v, v) which is a subdiagram of D such that the vertex v' diagonally opposite to v belongs to the boundary of D. Its sides are two opposite A-chains and two opposite B-chains with length ≥ 1. Note that their labels are words in a group alphabet of $G \in \mathcal{T}$ that represent geodesics in the Cayley graph $\Gamma(G, S)$. Let δ_v be half the length of its perimeter. We are looking for a constant δ_G such that $\sup\{\delta_v \mid v \in \dot{D}, (D, \phi, p, v_0)$ is an ARDD over $P_G\} \leq \delta_G$. It is clear that the orientation of the edges in D does not affect such δ_G. That is why to every 4-cell we will associate an unoriented unit square with labeled ("colored") sides, and to every D_v some "tiled" rectangle. By doing this we will enter the tiling problem originally stated by H. Wang, with the connection to a class of $\forall\exists\forall$-formulas in propositional calculus (see [14]).

3.1.1. Solvable periodic and aperiodic sets of protiles. Suppose that we are given a *finite* set T of unit squares (Wang's prototiles) with colored edges, placed with their edges horizontally and vertically. We will be interested in tiling the plane with copies (tiles) of these prototiles such that their vertices are placed on the vertices of the unit lattice of points in the plane; colors on contiguous edges must match, and only translation of the prototiles is allowed. We say that the set T is *solvable* or admits a tiling of the plane if we can tile the plane with tiles from T in the above described way. Note that if T is a solvable set of prototiles and T' includes T, then T' is also solvable. Every solvable set of prototiles T will define a tiling $\tau(T)$ (not necessarily unique) and for every such tiling $\tau(T)$ the notion of the symmetry group of a tiling is naturally defined (see [7]). It is the group of all isometries σ of the plane which map the tiles of $\tau(T)$ onto the tiles of $\tau(T)$, preserving the colors of the corresponding edges. Since only translation of the prototiles is permitted, we may label the horizontal colors in T (that is, the colors of the horizontal edges of the prototiles in T) with even numbers or with labels from some finite set B, called the set of B-symbols and the vertical colors with odd numbers or from some finite set of A-symbols (see Figure 3.1).

If a given tiling $\tau(T)$ remains invariant under translations in two non-parallel directions, we say that $\tau(T)$ is *periodic*. In any such tiling, we can find a tiled rectangle of some size which repeats to cover the plane. If a tiling $\tau(T)$ is not periodic, then we call it *non-periodic*.

Given a set of prototiles T we say that T is *periodic* if it admits a periodic tiling. If T is solvable, and none of the tilings $\tau(T)$ is periodic, then T is an *aperiodic* set of prototiles. It is easy to construct a periodic set of prototiles, but the existence of an aperiodic set was a challenging problem, first solved by Berger

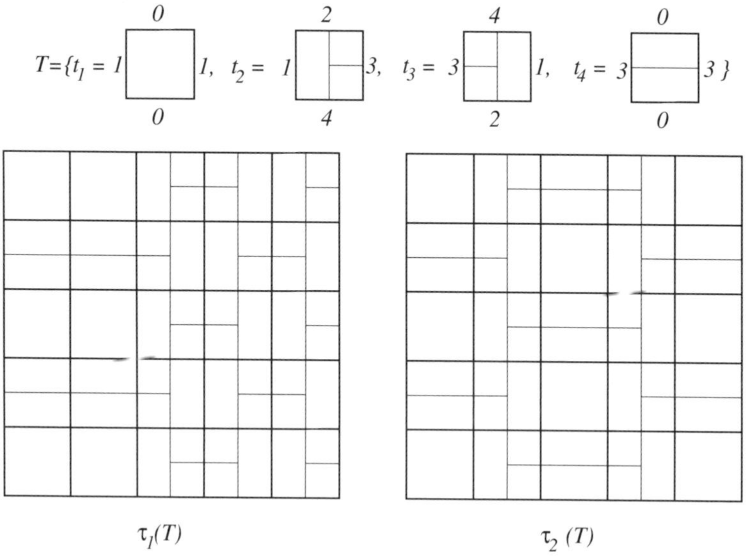

FIGURE 3.1. The figure illustrates portions of two different tilings $\tau_i(T)$ of the plane where $T = \{t_1, t_2, t_3, t_4\}$. Colors on the edges of t_i's are from the set $\{0, 1, 2, 3, 4\}$

(see [2]) who was able to produce an aperiodic set with $20,436$ prototiles. Today, this number has been reduced to 16 (i.e., the smallest known aperiodic set contains 16 prototiles, [7]). Berger's aperiodic set refuted H. Wang's conjecture that every solvable set of prototiles must be periodic. In doing so, he gave a negative answer to the tiling problem. The *tiling problem* is stated as follows: is there an algorithm or a standard procedure for deciding whether a given set of prototiles is solvable or not? The tiling problem is said to be *decidable* if there exists an algorithm which will yield a solution of it for **any** given set of prototiles in a finite number of steps.

An important theorem in establishing Berger's result is the following theorem due to H. Wang.

THEOREM 3.1.1. (see [7], [14]) *Let T be a given finite set of prototiles. If it is possible for arbitrary large values of n to assemble $n \times n$ blocks of tiles satisfying the color matching conditions, then T is solvable.*

For our purposes, we will restrict our attention to two subclasses of the class of all finite sets of prototiles. We will define the class $\mathcal{E}$ of all uniquely extendible sets of prototiles and the class $\mathcal{P}$ of all perfect sets of prototiles as follows.

Let T be a given set of prototiles. We will say that T is *perfect at the left upper corner* if no two prototiles from T have the same pair of colors on their left vertical and upper horizontal edges. The notion of being perfect at the right upper (lower) corner can be defined similarly. We say that T is *perfect* if it is perfect at each corner.

EXAMPLE 3.1.1. Let $T = \{ t_1 = \boxed{5 \; \overset{0}{} \; 1 \atop 2}, \; t_2 = \boxed{1 \; \overset{2}{} \; 3 \atop 0}, \; t_3 = \boxed{3 \; \overset{0}{} \; 5 \atop 2} \}$. Then T is an example of a perfect set of prototiles.

The class of all perfect sets of prototiles will be denoted by $\mathcal{P}$.

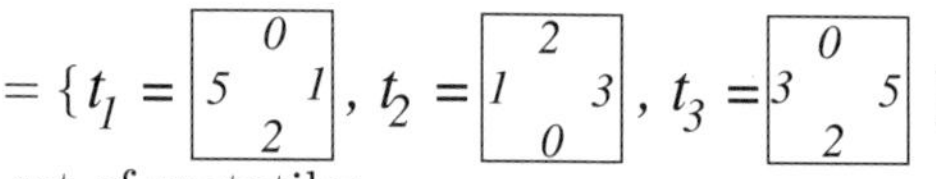

3.1.2. Extendability questions. Let T be a given set of prototiles, not necessarily solvable or perfect, and let $t_1, t_2 \in T$. We say that t_1 and t_2 form a *horizontal* (*vertical*) *pair* (t_1, t_2), $(\binom{t_1}{t_2})$ if they can be glued along a vertical (horizontal) edge, with t_1 being on the left (top) of t_2. A horizontal pair (t_1, t_2) of prototiles is said to be *vertically extendible* if there are horizontal pairs (t'_1, t'_2) and (t''_1, t''_2) such that the prototiles t'_i, t_i and t''_i, $(i = 1, 2)$ form vertical pairs $\binom{t'_i}{t_i}$ and $\binom{t_i}{t''_i}$, $(i = 1, 2)$. Similarly, we can define when a vertical pair $\binom{t_i}{t_j}$ is horizontally extendible. If horizontal pairs (t'_1, t'_2) and (t''_1, t''_2) from above and below, happen to be unique we say that the horizontal pair (t_1, t_2) is *uniquely extendible.* The set of prototiles T is said to be *uniquely extendible* if every extendible pair is uniquely extendible.

The class of all uniquely extendible sets of prototiles will be denoted by $\mathcal{E}$. The set T in Example 7.3 is an example of a solvable and uniquely extendible set of prototiles. Looking at the class of all prototiles it is natural to introduce an equivalence relation $\sim$ as follows:

$$T_1 \sim T_2 \text{ iff } \begin{cases} T_1 \text{ and } T_2 \text{ are not solvable} \\ T_1 \text{ and } T_2 \text{ are solvable and both periodic} \\ T_1 \text{ and } T_2 \text{ are solvable and both aperiodic} \end{cases}$$

For example, the set $T_0 = \{$ 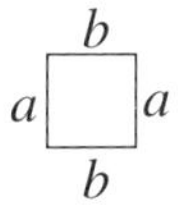

$\}$ is equivalent to the set T from Example 3.1.1.

3.1.3. Groups associated with sets of prototiles. Let $T = \{t_1, \ldots, t_n\}$ be a given set of prototiles, not necessarily solvable or perfect, and let I (J) be the set of all horizontal (vertical) pairs of prototiles in T. We define the set K as the set of all possible quadruples of prototiles, i.e.,

$$K \stackrel{\text{def}}{=} \left\{ \begin{pmatrix} t_i & t_j \\ t_k & t_l \end{pmatrix} \mid (t_k, t_l), (t_i, t_j) \in I, \begin{pmatrix} t_i \\ t_k \end{pmatrix}, \begin{pmatrix} t_j \\ t_l \end{pmatrix} \in J \right\}.$$

Let $A = \{a_i, \alpha_i, a_k^j, \alpha_k^j \mid i = 1, \ldots, n, \binom{t_j}{t_k} \in J\}$ and $B = \{b_l, \beta_l, b_{sq}, \beta_{sq} \mid l = 1, \ldots, n, (t_s, t_q) \in I\}$ be two finite sets of "A-colors" and "B-colors" respectively. To every finite set of prototiles T we can associate a new set of prototiles $S(T)$, called the *shuffle* of T in the following way:

For every $i = 1, \ldots, n$, there will be a prototile t'_i in $S(T)$ with the top color b_i, the left color α_i, the right color a_i and the bottom color β_i. For every horizontal pair (t_i, t_j) in I there will be a prototile t_{ij} in $S(T)$ with the top color b_{ij}, the bottom color β_{ij}, the left color a_i and the right color α_j. For every vertical pair $\binom{t_i}{t_k}$ in J there will be a prototile t_k^i with the top color β_i, the bottom color b_k, the left color α_k^i and the right color a_k^i. And finally, for every element $\begin{pmatrix} t_i & t_j \\ t_k & t_l \end{pmatrix}$ in K there will be a prototile t_{kl}^{ij} in $S(T)$ with the top color β_{ij}, the bottom color b_{kl}, the left color a_k^i and the right color α_l^j. Note that in this case there will be a 3×3 tiled square with tiles from $S(T)$ such that t_{kl}^{ij} is the central prototile in this 3×3 square (see Figure 3.2). The idea is simple. We recolor the edges of the prototiles in T such that the new set of prototiles $S(T)$ preserves the tiling properties of T.

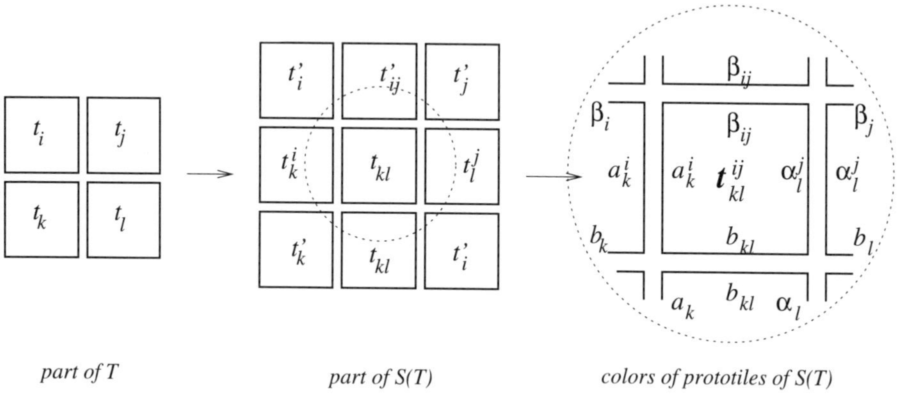

FIGURE 3.2

EXAMPLE 3.1.2. Let $T = \{t_1 = \begin{array}{|ccc|}\hline & 0 & \\ 1 & & 3 \\ & 0 & \\ \hline\end{array}\,,\ t_2 = \begin{array}{|ccc|}\hline & 2 & \\ 3 & & 1 \\ & 2 & \\ \hline\end{array}\ \}$. Obviously, T is periodic. $S(T)$ has 8 elements listed in Figure 3.3.

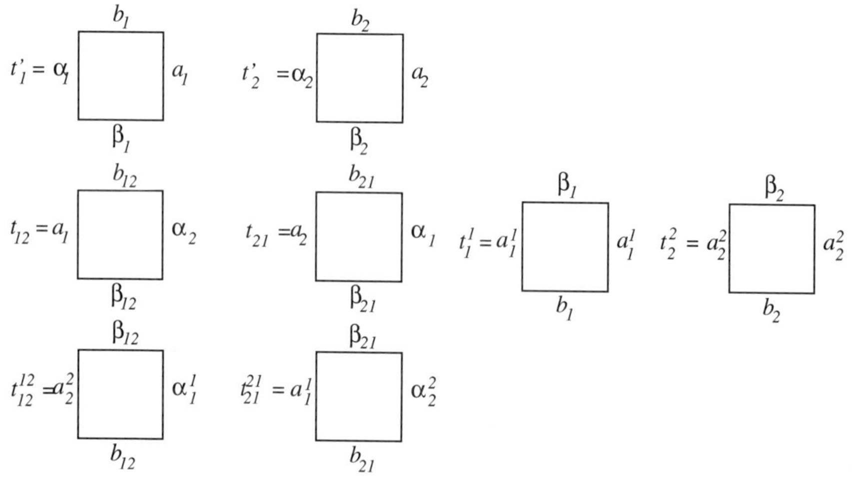

FIGURE 3.3

It follows from the definition of $S(T)$ that $T \sim S(T)$.

NOTE 3.1.1. Assume that $T \in \mathcal{E}$ and t_1, t_2 in T have the same pair of colors at the right upper vertex. If there exist prototiles t_i, t_j, t_k in T such that $\begin{pmatrix} t_i & t_j \\ t_1 & t_k \end{pmatrix} \in K$ then we have that $\begin{pmatrix} t_i & t_j \\ t_2 & t_k \end{pmatrix}$ is also in K. Therefore the vertical pair $\binom{t_j}{t_k}$ is not uniquely extendible, contrary to the assumption that $T \in \mathcal{E}$. So no such t_i, t_j and t_k exist and consequently if:

(i) T is solvable then t_1 and t_2 are not used in any of the tilings $\tau(T)$, i.e., $T \sim T \setminus \{t_1, t_2\}$;
(ii) T is not solvable then $T \setminus \{t_1, t_2\}$ is also not solvable.

In any case, for every $T \in \mathcal{E}$ we can find $T' \in \mathcal{P}$ such that $T \sim T'$. If $\mathcal{E}/_\sim$ is the set of all equivalence classes of uniquely extendible sets of prototiles then $\mathcal{E}/_\sim \subseteq \mathcal{P}/_\sim$. So, we can assume that every uniquely extendible set of prototiles is perfect.

DEFINITION 3.1.1. Let T be a given set of prototiles, not necessarily perfect. *The group $\mathcal{G}(T)$ associated to T* is determined by the following presentation $P_{\mathcal{G}(T)}$:

generators: one generator for every different color in $S(T)$.
relators: one relator $ba\beta^{-1}\alpha^{-1}$ for every prototile t in $S(T)$ with top color b, bottom color β, left color α and right color a.

EXAMPLE 3.1.3. Continuing Example 3.1.2 we will have the following presentation $P_{\mathcal{G}(T)}$ for the group $\mathcal{G}(T)$:

generators: $a_1, a_2, a_1^1, a_2^2, \alpha_1, \alpha_2, \alpha_1^1, \alpha_2^2, b_1, b_2, \beta_1, \beta_2, b_{12}, b_{21}, \beta_{12}, \beta_{21}$
relations: $b_1 a_1 = \alpha_1 \beta_1$, $b_2 a_2 = \alpha_2 \beta_2$ $b_{12}\alpha_2 = a_1 \beta_{12}$, $b_{21}\alpha_1 = a_2 \beta_{21}$,
$\beta_1 a_1^1 = \alpha_1^1 b_1$, $\beta_2 a_2^2 = \alpha_2^2 b_2$, $\beta_{12}\alpha_1^1 = a_2^2 b_{12}$, $\beta_{21}\alpha_2^2 = a_1^1 b_{21}$

By applying some elementary Tietze transformations, we can eliminate a_1, a_2, a_1^1, $a_2^2, \alpha_2, \alpha_2^2$ and obtain a new presentation for $\mathcal{G}(T)$:

generators: $a, \alpha_1, b, b_1, b_{12}, b_2, \beta_1, \beta_2, \beta_{12}, \beta_{21}$
relations: $ba = ab$

We see that $\mathcal{G}(T) \cong (\mathbb{Z} \times \mathbb{Z}) * \underbrace{\mathbb{Z} * \cdots * \mathbb{Z}}_{8 \text{ times}}$. Note that $\mathbb{Z} \times \mathbb{Z}$ is a subgroup of $\mathcal{G}(T)$.

PROPOSITION 3.1.1. *If T is a perfect set of prototiles, $\mathcal{G}(T) \in \mathcal{T}$.*

PROOF. There are four types of relators in $P_{\mathcal{G}(T)}$: relators r_i' associated with the prototiles t_i' in $S(T)$, relators r_{ij} and r_j^i associated with the prototiles t_{ij} and t_j^i respectively, and relators r_{kl}^{ij} associated with the prototiles t_{kl}^{ij}. Every relator has length 4 and has the form $ba\beta^{-1}\alpha^{-1}$ for some $b, \beta \in B$ and $a, \alpha \in A$. The only thing to check is the condition (C) in the definition of the class $\mathcal{T}$. None of the relators r_i' have common symbols and they can have at most one common symbol with a relator of the form r_{ij} or r_j^i. Similarly, a relator of the form r_{ij} (r_j^i) can have at most one common symbol with some of the relators r_{ij}, r_j^i and r_{kl}^{ij}. So, let us assume that two relators r_{kl}^{ij} and r_{st}^{pq} have the same prefixes of length two. Let $r_{kl}^{ij} \equiv \beta_{ij}\alpha_l^j b_{kl}^{-1}(a_k^i)^{-1}$ and $r_{st}^{pq} \equiv \beta_{pq}\alpha_t^q b_{st}^{-1}(a_s^p)^{-1}$. By the assumption $i = p$, $q = j$ and $t = l$. Because T is perfect (at the right upper corner), it follows that $s = k$ i.e., $r_{kl}^{ij} \equiv r_{st}^{pq}$. □

THEOREM 3.1.2. *Let $T \in \mathcal{E}$. $\mathcal{G}(T)$ is hyperbolic iff T is not solvable.*

PROOF. $\Leftarrow$ Let $T = \{t_1, \ldots, t_n\}$ be given. As noted in Proposition 3.1.1, we can assume that T is perfect. There are four types of relators in $P_{\mathcal{G}(T)}$. "Original" relators $r_i' \equiv b_i a_i \beta_i^{-1}\alpha_i^{-1}$, $(i = 1, \ldots, n)$, associated with the prototiles t_i' in $S(T)$, "horizontal arm" relators $r_{ij} \equiv b_{ij}\alpha_j\beta_{ij}^{-1}a_i^{-1}$, associated with the prototiles t_{ij} in $S(T)$, "vertical arm" relators $r_j^i \equiv \beta_i a_j^i b_j^{-1}(\alpha_j^i)^{-1}$, associated with the prototiles t_j^i in $S(T)$ and "cross" relators $r_{kl}^{ij} \equiv \beta_{ij}\alpha_l^j b_{kl}^{-1}(a_k^i)^{-1}$ associated with the prototiles t_{kl}^{ij} in $S(T)$.

By Theorem 3.1.1, since T is not solvable, there is a natural number $n_0 > 1$ such that for any $m \times m$ tiled square, tiled with prototiles from T, $m < n_0$. Because $\mathcal{G}(T) \in \mathcal{T}$ it is sufficient to show that the thickness of the geodesic digons in $\Gamma(\mathcal{G}(T), A \cup B)$ is less than some fixed number which depends on n_0. We will show that this number is $2(2n_0 - 1)$. Let (D, ϕ, p, v_0) be a disk diagram having two geodesics on its boundary and let C be an interior cell in D. We know that D is a regular ARDD, so we will estimate the size of the biggest rectangular disk subdiagram D_v of D $v \in \dot{D}^{(o)}$. The label of the boundary cycle of C, $\phi(\dot{C})$, can represent (mod cyclic permutation) one of the relators r'_i, r_{ij}, r^i_j, r^{ij}_{kl}. So we can say that $\phi(\dot{C})$ (mod cyclic permutation) can be an original relator, a vertical arm, a horizontal arm or a cross relator respectively. There are four possibilities for $\phi(\dot{C})$. If

(1) $\phi(\dot{C})$ is an original relator, the labels of the four cells C_l, C_r, C_u and C_d surrounding C from the left, right, top and bottom, respectively, must be arm relators and the labels of the four cells C_{lu}, C_{ru}, C_{ld} and C_{rd}, having only the right lower vertex, the left lower vertex, the right upper vertex and the left upper vertex respectively in common with C, must be the cross relators. (If a vertex of C is on the boundary of D, some of these cells will not exist.) So the "cell neighborhood" of C will look like the one shown in Figure 3.4.

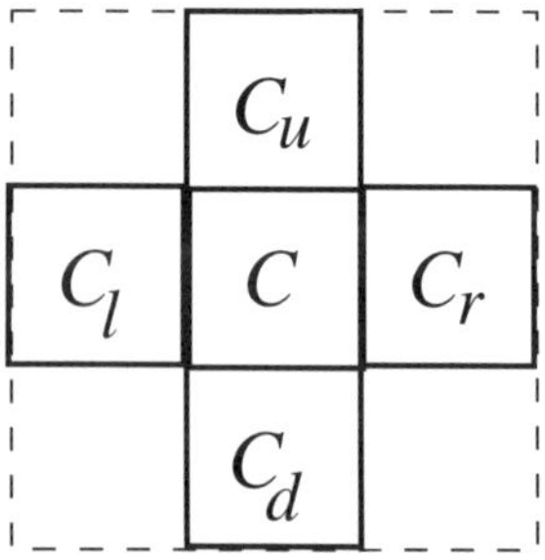

FIGURE 3.4. This figure shows a subdiagram of (D, ϕ, p, v_0). The orientation of the edges of C determines the orientation of the other edges.

Since the horizontal edges of C are labeled with B symbols there are precisely four ways in which the edges of the 4-cell C might be oriented. Once we know which of these appears in C, the orientations of all solid-line edges in Figure 3.4 are determined.

(2) $\phi(\dot{C})$ is a horizontal arm relator, there are two possibilities for C_l. Let $\phi(\dot{C}_l)$ also be an arm relator and let us specify the labels on $\dot{C}$ and $\dot{C}_l$ as $r_{ij} \equiv b_{ij}\alpha_j\beta_{ij}^{-1}a_i^{-1}$ and $r_{ik} \equiv b_{ik}\alpha_k\beta_{ik}^{-1}a_i^{-1}$ respectively. $\phi(\dot{C}_u)$ must be a cross relator and the orientation of the vertical edges of C_u depends on the orientation of the vertical edges of C. In Figure 3.5. the orientation is downwards.

The cell C_u and the cell C_{lu} (if such exists) represent cross relators and have a common vertical edge. Because $T \in \mathcal{E}$ this is possible iff the labels on the boundaries coincide (mod cyclic perm.), i.e., (D, ϕ, p, v_0) is not a reduced

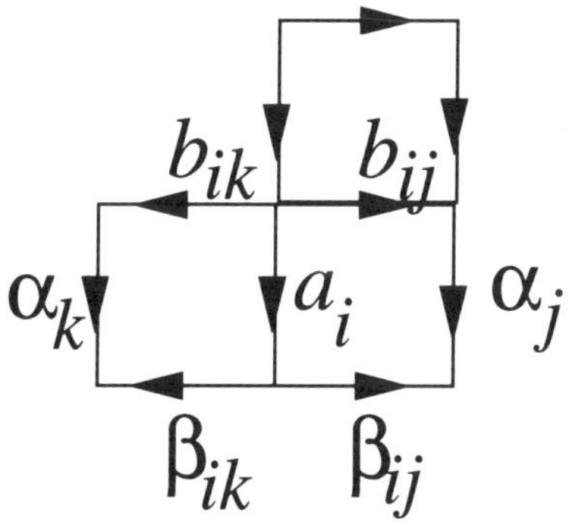

FIGURE 3.5

disk diagram. So, if no vertex of the vertical arm cell is on the boundary of D, its left and right neighbors must be original cells.

(3), (4) If $\phi(\dot{C})$ is a horizontal arm relator, arguments similar to these in (2) show that $\phi(\dot{C}_u)$ and $\phi(\dot{C}_d)$ are original relators. If $\phi(\dot{C})$ is a cross relator, the same arguments as in (1) show that $\phi(\dot{C}_u)$, $\phi(\dot{C}_d)$ are vertical arm relators and $\phi(\dot{C}_l)$, $\phi(\dot{C}_r)$ are horizontal arm relators.

Therefore, there is no RDD D_v, a subdiagram of D, with length of its side greater than $2n_0 - 1$. Thus $\mathcal{G}(T)$ is hyperbolic with hyperbolicity constant $\delta_G = 2(2n_0 - 1)$ (see the end of 2.2.1).

$\Rightarrow$ If T is solvable, by Theorem 3.1.1, for any $m \in \mathbb{Z}^+$ there is a tiled square of size $m \times m$, tiled with prototiles from T. But every such square gives a tiled $(2m-1) \times (2m-1)$ square, tiled with prototiles from $S(T)$, i.e., there is no bound on the length of the side of D_v for a RDD (D, ϕ, p, v_0) when v is a corner of D. Therefore $\mathcal{G}(T)$ is not hyperbolic. $\square$

EXAMPLE 3.1.4. Unfortunately, for this class of groups, the result of the previous proposition is the best possible in the sense that there exists a finite set of prototiles $T = \{t_1, \ldots, t_{16}\}$ which is perfect (but not uniquely extendible) such that $\mathcal{G}(T)$ is not hyperbolic. Instead of giving the colors on each of the prototiles t_i, $(i = 1, \ldots, 16)$, we will specify the set K of all possible quadruples $\begin{pmatrix} t_i & t_j \\ t_k & t_l \end{pmatrix}$ $\left(= \begin{pmatrix} i & j \\ k & l \end{pmatrix}\right.$ for short) in Figure 3.6.

We read Figure 3.6 in the following way. The tiled 2×2 square

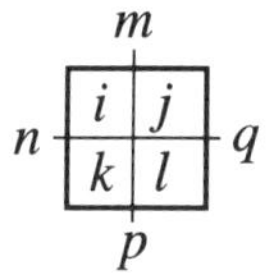

gives the following information:

The right vertical color of the prototile $t_i(t_k)$ is $a_m(a_p)$ and it is equal to the left vertical color of the prototile $t_j(t_l)$. Similarly, the bottom horizontal color of the prototile $t_i(t_j)$ is $b_n(b_q)$ and it is equal to the top horizontal color of the prototile $t_k(t_l)$. The left vertical (top horizontal) color of t_i is defined such that it does not coincide with any other vertical (horizontal) color of the prototiles in T. Similarly, we define the right vertical color of $t_j(t_l)$ or the top (bottom) horizontal color of $t_j(t_l)$. Clearly no square of side greater than two can be tiled, so this set of prototiles is not solvable.

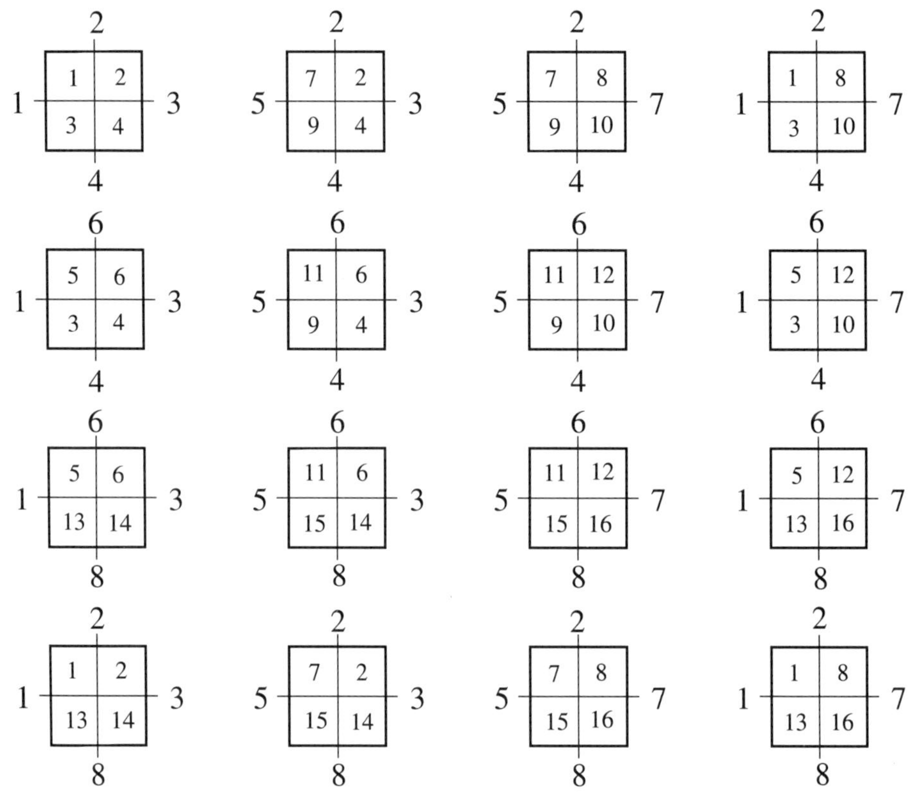

FIGURE 3.6

The group $\mathcal{G}(T)$ will have 104 generators ($|A| = |B| = 52$), 64 defining relators, which can easily be deduced from the Definition 3.1.1. It is not too difficult to see that the C-geodesics $U \equiv b_{12}b_{72}^{-1}b_{78}b_{18}^{-1}$ and $V \equiv a_3^1(a_3^5)^{-1}a_{13}^5(a_{13}^1)^{-1}$ commute, therefore generate free abelian subgroup of $\mathcal{G}(T)$ of rank 2. A disk diagram associated to the commutator $[U, V]$ is given on Figure 3.7.

THEOREM 3.1.3. *If $T \in \mathcal{P}$ and T is periodic, then $\mathbb{Z} \times \mathbb{Z}$ is a subgroup of $\mathcal{G}(T)$.*

PROOF. If T is a perfect periodic set of prototiles, $S(T)$ is also perfect and periodic. Let Q_0 be a tiled square of size $n_0 \times n_0$ tiled with prototiles of $S(T)$ which yields a periodic tiling. Note that $n_0 \geq 2$ (n_0 is an even number). In order to simplify the notation we take that $n_0 = 4$. The general case follows the same line of argument. Let t_i, t_j, t_k and t_l in T be such that there is a prototile t_{kl}^{ij} in $S(T)$. The label on the top (bottom) horizontal side of Q_0 reading from left to right, is $U \equiv \beta_k\beta_{kl}\beta_l\beta_{lk}$ and the label on the left (right) vertical side of Q_0, reading from top to bottom, is $V \equiv \alpha_i^k\alpha_i\alpha_k^i\alpha_k$. We want to show that the elements U and V generate a free abelian subgroup of rank 2. Note that U and V are C-geodesics and there is a RDD $\left(D_0, \phi, pq(p')^{-1}(q')^{-1}, v_0\right)$ such that $\phi(p) \equiv \phi(p') \equiv U$ and $\phi(q) \equiv \phi(q') \equiv V$. So $UVU^{-1}V^{-1}$ is a relator in $P_{\mathcal{G}(T)}$. If there is another relator $F(U, V)$ involving U and V, then using the relation $UV = VU$, $F(U, V)$ can be reduced to the form U^pV^q for some $p, q \in \mathbb{Z}$. But this contradicts Corollary 2.2.3 because U is a non-empty B-word and V is a non-empty A-word. By the same

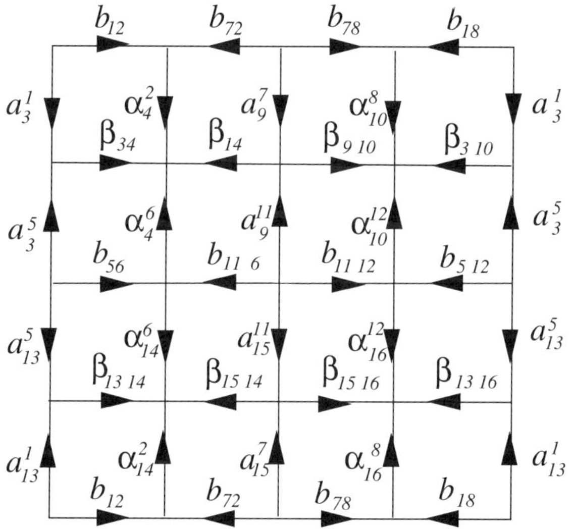

FIGURE 3.7

corollary (D_0 has at least one 4-cell) no relator of the form U^k (V^k) is possible. So U and V have infinite order in $\mathcal{G}(T)$. Therefore they generate a free abelian subgroup of rank two. □

3.2. Tilings Associated With a Given Presentation

3.2.1. Tilings and groups in $\mathcal{T}$. Let $P_G = \langle S \mid \mathcal{R} \rangle$ be a given presentation of $G \in \mathcal{T}$. To this presentation we will associate a finite set of Wang prototiles $\mathbb{T}(P_G)$, reducing the question of hyperbolicity of G to a question of solvability of $\mathbb{T}(P_G)$. The construction of $\mathbb{T}(P_G)$ in some respect imitates the construction of the shuffle $S(T)$, but there are some minor subtleties regarding the relators in $\mathcal{R}_2$. Basically, to every rectangular disk diagram of the form $D_v \subseteq D$, $v \in \dot{D}^{(o)}$, for an ARDD $(D, \phi, \dots)$ we will associate a tiled square with length of its side equal to 2(length of the side of D_v)-1.

Let $B(\bar{\mathcal{R}}_{4*}) = \{r_1, \dots, r_n\}$. For every element r_i in $B(\bar{\mathcal{R}}_{4*})$, $(i = 1, \dots, n)$, we will introduce a prototile $t(i)$ in $\mathbb{T}(P_G)$ with the top color $b(i)$, the bottom color $\beta(i)$, the left color $\alpha(i)$ and the right color $a(i)$. We see that if $r_i \not\equiv r_j$ then $t(i)$ and $t(j)$ have no common colors. Let $r_i \equiv b_i^{\epsilon_i} a^{\mu} \beta_i^{\eta_i} \alpha_i^{\xi_i}$ and $r_j \equiv b_j^{\epsilon_j} a_j^{\mu_j} \beta_j^{\nu_j} a^{-\mu}$ be two elements in $B(\bar{\mathcal{R}}_{4*})$ not necessarily distinct. We will introduce a prototile $t(i,j)$ having $b(i,j)$ as its top color, $\beta(i,j)$ as its bottom color, $a(i)$ as its left color and $\alpha(j)$ as its right color iff each of the words $b_i^{\epsilon_i} b_j^{\epsilon_j}$ and $\beta_j^{\nu_j} \beta_i^{\nu_i}$ are C-geodesics. Similarly, let $r_k \equiv b_k^{\epsilon_k} a_k^{\mu_k} \beta^{-\nu} \alpha_k^{\epsilon_k}$ and $r_l \equiv \beta^{\nu} a_l^{\mu_l} \beta_l^{\nu_l} \alpha_l^{\xi_l}$ be two elements in $B(\bar{\mathcal{R}}_{4*})$ not necessarily distinct. We will introduce a prototile $t\binom{k}{l}$ having $\beta(k)$ as its top color, $b(l)$ as its bottom color, $\alpha\binom{k}{l}$ as its left color and $a\binom{k}{l}$ as its right color iff each of the words $a_k^{\mu_k} a_l^{\mu_l}$ and $\alpha_l^{\xi_l} \alpha_k^{\xi_k}$ are C-geodesics. Finally, for every four relators r_i, r_j, r_k and r_l in $B(\bar{\mathcal{R}}_{4*})$ such that $t(i,j)$, $t(k,l)$, $t\binom{i}{k}$ and $t\binom{j}{l}$ exist, we introduce a prototile $t\binom{ij}{kl}$ having the top color $\beta(i,j)$, the bottom color $b(k,l)$ the left color $a\binom{i}{k}$ and the right color $\alpha\binom{j}{l}$.

EXAMPLE 3.2.1. Let $P_G = \langle a, b \mid baba, b^2 \rangle$ be a presentation of $D_\infty \in \mathcal{T}$, the infinite dihedral group as in Example 3 on page 8. Then

$$B(\bar{\mathcal{R}}_{4*}) = \{r_1 \equiv baba, r_2 \equiv b^{-1}aba, r_3 \equiv bab^{-1}a, r_4 \equiv b^{-1}ab^{-1}a,$$
$$r_5 \equiv b^{-1}a^{-1}b^{-1}a^{-1}, r_6 \equiv ba^{-1}b^{-1}a^{-1}, r_7 \equiv b^{-1}a^{-1}ba^{-1}, r_8 \equiv ba^{-1}ba^{-1}\}$$

We can see that no prototiles of type $t(i, j)$ or $t\binom{ij}{kl}$ exist. So $\mathbb{T}(P_G)$ will contain the following prototiles

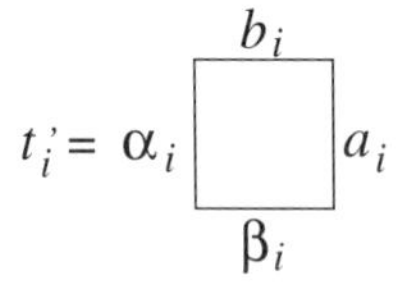

$$t_i' = \alpha_i \quad \boxed{\begin{array}{c} b_i \\ \\ \beta_i \end{array}} a_i \qquad \text{for } i = 1, \ldots, 8$$

$$t_j^i = \alpha_j^i \quad \boxed{\begin{array}{c} \beta_i \\ \\ b_j \end{array}} a_j^i \qquad \text{for } \binom{i}{j} \in \{\binom{1}{2}, \binom{1}{4}, \binom{2}{2}, \binom{2}{4}, \binom{3}{1}, \binom{3}{3}, \binom{4}{1}, \binom{4}{3}, \binom{5}{6}, \binom{5}{8}, \binom{6}{6}, \binom{6}{8}, \binom{7}{5}, \binom{7}{7}, \binom{8}{5}, \binom{8}{7}\}$$

Note that $\mathbb{T}(P_G)$ is not solvable and that $\mathbb{Z} \times \mathbb{Z}$ is not a subgroup of G. Actually we can find all simple geodesic triangles in $\Gamma(G, A \cup B)$ by noting that there is no RDD such that the length of the smaller of its sides is greater than 1. So the set of all RDD's coincides with the set of all ARDD's and to every RDD we can associate a path of finite length in one of the two directed graphs in Figure 3.8 where the vertices 1 through 8 represent the relators r_1 through r_8 and two vertices i and j are connected with an oriented edge from i to j iff there is a prototile $t\binom{i}{j}$.

$$(i,j,k,l) \in \{ (1,2,3,4), (5,6,7,8) \}$$

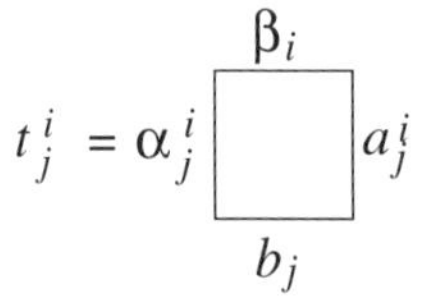

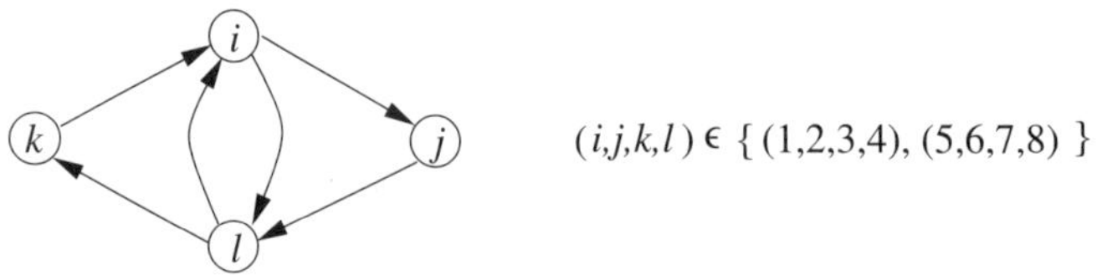

FIGURE 3.8

PROPOSITION 3.2.1. *If P_G is a presentation of $G \in \mathcal{T}$ then $\mathbb{T}(P_G)$ is a perfect set of prototiles.*

PROOF. Following the definition of $\mathbb{T}(P_G)$, $G \in \mathcal{T}$ we can see that there are four types of prototiles. The first type are the prototiles $t(i)$ associated with the relators r_i in $B(\bar{\mathcal{R}}_{4*})$. We can call them original prototiles. The second and third type are the prototiles $t(i, j)$ and $t\binom{k}{l}$ associated with the pairs of words in $B(\bar{\mathcal{R}}_{4*})$. We can call them vertical and horizontal arms, respectively. The fourth type are prototiles $t\binom{ij}{kl}$, so called crosses. It is easy to see that no two original or vertical (horizontal) arms have the same pair of colors at the same corner. If $t\binom{ij}{kl}$ and $t\binom{pq}{mn}$ have the same colors at the lower left corner, i.e., if $a\binom{i}{k} \equiv a\binom{p}{m}$ and $b(k,l) \equiv b(m,n)$ then $i = p$, $k = m$ and $l = n$. Assume that $j \neq q$, i.e., $r_j \not\equiv r_q$. If $r_q \equiv b_q^{\epsilon_q} a_q^{\mu_q} \beta_q^{\nu_q} \alpha_q^{\xi_q}$ and $r_j \equiv b_j^{\epsilon_j} a_j^{\mu_j} \beta_j^{\nu_j} \alpha_j^{\xi_j}$ then $\beta_q \equiv \beta_j$, $\nu_q = \nu_j$; $\alpha_q \equiv \alpha_j$, $\xi_q = \xi_j$ because both quadruples (r_i, r_j, r_k, r_l) and (r_i, r_q, r_k, r_l) satisfy the condition which leads to a construction of $t\binom{ij}{kl}$ and $t\binom{iq}{kl}$. By the condition (C) from the definition of $\mathcal{T}$ in Section 1.2, $b_j \equiv b_q$, $\epsilon_j = \epsilon_q$; $a_j \equiv a_q$, $\mu_j = \mu_q$, i.e., $j = q$. □

THEOREM 3.2.1. *$G \in \mathcal{T}$ is hyperbolic iff $\mathbb{T}(P_G)$ is not solvable.*

PROOF. $\Rightarrow$: Assume that δ_G is a hyperbolicity constant for G. Let Δxyz be a C-geodesic triangle in $\Gamma(G, A \cup B)$. Then the thickness of every r-component of Δ is $\leq \delta_G$, i.e., there is no $D_\Box$ with length of its side greater than $\delta_G/2$. Knowing that to every $D_\Box$ with length of its side m we can associate a tiled square, tiled with prototiles from $\mathbb{T}(P_G)$ such that the length of its side is $2m-1$, it follows that no such square can have a side of length greater than $\delta_G - 1$. By Theorem 3.1.1 $\mathbb{T}(P_G)$ is not solvable.

$\Leftarrow$: Assume that $\mathbb{T}(P_G)$ is not solvable. Then there exists an $n_0 \in \mathbb{Z}^+$ such that for every $m \times m$ tiled square, tiled with prototiles from $\mathbb{T}(P_G)$, $m \leq n_0$. This implies that the thickness of every geodesic digon in $\Gamma(G, A \cup B)$ is $\leq 2n_0$. Let Δxyz be a geodesic triangle in $\Gamma(G, A \cup B)$ and $\Delta' xyz$ be the C-geodesic triangle with the same set of vertices as Δxyz and with geodesic sides p'_{xy}, p'_{yz} and p'_{zx}. Let $v_0 \in p'_{zx}$ and let v'_0 be the diagonally opposite vertex of v_0 in $(D_\Box, \phi, p_\Box, v_0)$ associated with v_0. Note that if $\Delta' xyz$ contains an interior vertex v_6 such that $d_4(v_6) = 6$ then $v'_0 \in p_1 \cup p_2 \cup p_3$ (see 2.2.1). Then $d(v_0, p'_{xy} \cup p'_{yz})$, the distance between v_0 and $p'_{xy} \cup p'_{yz}$ is $\leq 4n_0$. But for any $v \in p_{zx}$, $d(v, p'_{zx}) \leq 2n_0$. So, $d(v, p_{xz} \cup p_{yz}) \leq 8n_0$, i.e., G is hyperbolic with hyperbolicity constant bigger or equal to $8n_0$. □

THEOREM 3.2.2. *Let $G \in \mathcal{T}$. $\mathbb{Z} \times \mathbb{Z}$ is a subgroup of G iff $\mathbb{T}(P_G)$ is periodic.*

PROOF. The "only if" part follows from Theorem 3.1.2.

Let U and V be C-geodesic representatives of the elements $\bar{U}$ and $\bar{V}$ which generate $\mathbb{Z} \times \mathbb{Z}$. As in Proposition 6.20, we can assume that U and V have no common prefixes and suffixes. Having that $UVU^{-1}V^{-1}$ is a relator in P_G, by Theorem 2.2.4, there will be a disk diagram $(D, \phi, pq(p')^{-1}(q')^{-1}, v_0)$ over P_G such that $\phi(p) \equiv \phi(p') \equiv U$, $\phi(q) \equiv \phi(q') \equiv V$ and D is an RDD. But then there will be a tiled rectangle of size $(2|U|-1) \times (2|V|-1)$ with prototiles from $\mathbb{T}(P_G)$ with two-by-two opposite sides having the same labels, i.e., $\mathbb{T}(P_G)$ is periodic. □

EXAMPLE 3.2.2. Let $P_G = \langle a, \alpha, b, \beta \mid b\alpha\beta^{-1}a, \beta ab^{-1}\alpha^{-1}, a^2 \rangle$ be a presentation of G in $\mathcal{T}$. It is not difficult to show that P_G is a presentation of $(\mathbb{Z} \times \mathbb{Z}) * \mathbb{Z}_2$. (Eliminating α from the defining relators we will get the new presentation $P'_G = \langle a, b, \beta \mid ab\beta a = \beta ab, a^2 = 1 \rangle$. Introducing new generators $x = ab$ and $y = \beta a$ and eliminating b and β we get $P''_G = \langle a, x, y \mid xy = yx, a^2 = 1 \rangle$.)

Instead of listing all the elements of $B(\bar{\mathcal{R}}_{4*})$ ($|B(\bar{\mathcal{R}}_{4*})| = 16$ and the set of all prototiles of the form $t(i,j)$ or $t\binom{k}{l}$ has 80 elements) we will label only two of them which will produce the periodicity of $\mathbb{T}(P_G)$. Let $B(\bar{\mathcal{R}}_{4*}) = \{r_1 \equiv b\alpha\beta^{-1}a^{-1}, r_2 \equiv \beta ab^{-1}\alpha^{-1}, \dots\}$. Then $\mathbb{T}(P_G)$ will contain the subset $\mathbb{T}_0 \sim \mathbb{T}(P_G)$ consisting of the following prototiles (each given with the corresponding symbol):

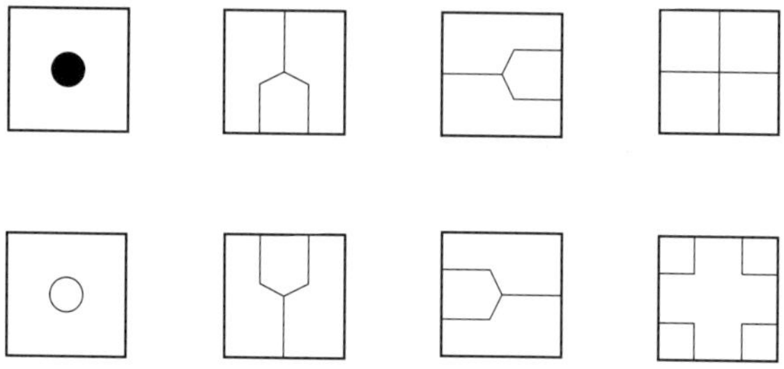

FIGURE 3.9. Prototiles

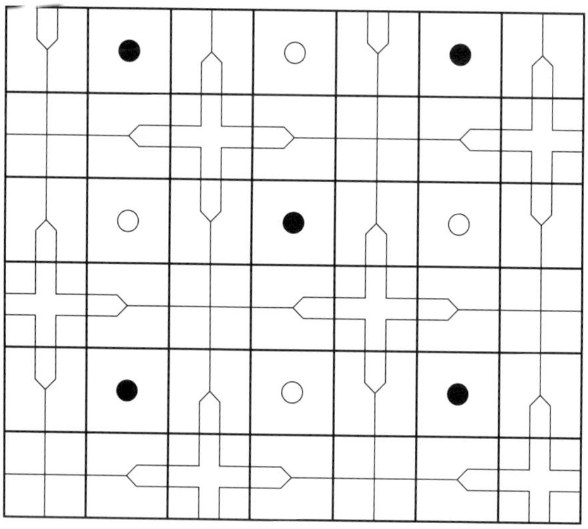

FIGURE 3.10

Using the symbolic presentation of the prototiles $t(i), t(k,l), \ldots$, part of the periodic tiling $\tau(\mathbb{T}_0)$ is presented in Figure 3.10.

Bibliography

1. J. M. Alonso, T. Brady, D. Cooper, V. Ferlini, M. Lustig, M. Mihalik, M. Shapiro, and H. Short, *Notes on word hyperbolic groups*, Group Theory from a Geometrical Viewpoint (E. Ghys, A. Haefliger, and A. Verjovsky, eds.), World Scientific, River Edge, NJ, 1991, pp. 3–64.
2. R. Berger, *The undecidability of the domino problem*, Memoirs of the AMS **66** (1966), 72 p.
3. D.B.A. Epstein, et al., *Word processing in groups*, Jones and Bartlet, Boston, 1992.
4. S. M. Gersten and H. B. Short, *Small cancellation theory and automatic groups*, Invent. Math. **102** (1990), 305–334.
5. ______, *Rational subgroups of biautomatic groups*, Annals of Mathematics **134** (1991), 125–158.
6. M. Gromov, *Hyperbolic groups*, Essays in Group Theory (S. M. Gersten, ed.), Springer-Verlag, 1987, pp. 75–265.
7. B. Grünbaum and G. S. Shepard, *Tilings and patterns*, W. H. Freeman & Company, New York, 1989, An introduction.
8. J. Kari, *Deterministic and aperiodic tiles*, manuscript, August 1994.
9. R. Lyndon and P. Schupp, *Combinatorial group theory*, Springer-Verlag, 1977.
10. W. Magnus, A. Karrass, and D. Solitar, *Combinatorial group theory*, Wiley, 1966.
11. A. Yu. Ol'shanskii, *Geometry of defining relations in groups*, Kluwer Academic Publishers, 1991.
12. A.Yu. Ol'shanskii, *Almost every group is hyperbolic*, Int. J. of Algebra and Comp. **2** (1992), 1–19.
13. R. Strebel, *Small cancellation groups*, Sur les Groupes Hyperboliques d'apres Mikhael Gromov (E. Ghys and P. de la Harpe, eds.), Progress in Math., vol. 83, Birkhäuser, Boston, 1990, pp. 227–273.
14. H. Wang, *Notes on a class of tiling problems*, Fundamenta Mathematicae **82** (1975), 295–334.

Editorial Information

To be published in the *Memoirs*, a paper must be correct, new, nontrivial, and significant. Further, it must be well written and of interest to a substantial number of mathematicians. Piecemeal results, such as an inconclusive step toward an unproved major theorem or a minor variation on a known result, are in general not acceptable for publication. Papers appearing in *Memoirs* are generally longer than those appearing in *Transactions*, which shares the same editorial committee.

As of May 31, 2001, the backlog for this journal was approximately 7 volumes. This estimate is the result of dividing the number of manuscripts for this journal in the Providence office that have not yet gone to the printer on the above date by the average number of monographs per volume over the previous twelve months, reduced by the number of volumes published in four months (the time necessary for preparing a volume for the printer). (There are 6 volumes per year, each containing at least 4 numbers.)

A Consent to Publish and Copyright Agreement is required before a paper will be published in the *Memoirs*. After a paper is accepted for publication, the Providence office will send a Consent to Publish and Copyright Agreement to all authors of the paper. By submitting a paper to the *Memoirs*, authors certify that the results have not been submitted to nor are they under consideration for publication by another journal, conference proceedings, or similar publication.

Information for Authors

Memoirs are printed from camera copy fully prepared by the author. This means that the finished book will look exactly like the copy submitted.

The paper must contain a *descriptive title* and an *abstract* that summarizes the article in language suitable for workers in the general field (algebra, analysis, etc.). The *descriptive title* should be short, but informative; useless or vague phrases such as "some remarks about" or "concerning" should be avoided. The *abstract* should be at least one complete sentence, and at most 300 words. Included with the footnotes to the paper should be the 2000 *Mathematics Subject Classification* representing the primary and secondary subjects of the article. The classifications are accessible from `www.ams.org/msc/`. The list of classifications is also available in print starting with the 1999 annual index of *Mathematical Reviews*. The Mathematics Subject Classification footnote may be followed by a list of *key words and phrases* describing the subject matter of the article and taken from it. Journal abbreviations used in bibliographies are listed in the latest *Mathematical Reviews* annual index. The series abbreviations are also accessible from `www.ams.org/publications/`. To help in preparing and verifying references, the AMS offers MR Lookup, a Reference Tool for Linking, at `www.ams.org/mrlookup/`. When the manuscript is submitted, authors should supply the editor with electronic addresses if available. These will be printed after the postal address at the end of the article.

Electronically prepared manuscripts. The AMS encourages electronically prepared manuscripts, with a strong preference for $\mathcal{A}\mathcal{M}\mathcal{S}$-LaTeX. To this end, the Society has prepared $\mathcal{A}\mathcal{M}\mathcal{S}$-LaTeX author packages for each AMS publication. Author packages include instructions for preparing electronic manuscripts, the *AMS Author Handbook*, samples, and a style file that generates the particular design specifications of that publication series. Though $\mathcal{A}\mathcal{M}\mathcal{S}$-LaTeX is the highly preferred format of TeX, author packages are also available in $\mathcal{A}\mathcal{M}\mathcal{S}$-TeX.

Authors may retrieve an author package from e-MATH starting from `www.ams.org/tex/` or via FTP to `ftp.ams.org` (login as `anonymous`, enter username as password, and type `cd pub/author-info`). The *AMS Author Handbook* and the *Instruction Manual* are available in PDF format following the author packages link from `www.ams.org/tex/`. The author package can be obtained free of charge by sending email to `pub@ams.org` (Internet) or from the Publication Division, American Mathematical Society, P.O. Box 6248, Providence, RI 02940-6248. When requesting an author package, please specify $\mathcal{A}\mathcal{M}\mathcal{S}$-LaTeX or $\mathcal{A}\mathcal{M}\mathcal{S}$-TeX, Macintosh or IBM (3.5) format, and the publication in which your paper will appear. Please be sure to include your complete mailing address.

Sending electronic files. After acceptance, the source file(s) should be sent to the Providence office (this includes any TeX source file, any graphics files, and the DVI or PostScript file).

Before sending the source file, be sure you have proofread your paper carefully. The files you send must be the EXACT files used to generate the proof copy that was accepted for publication. For all publications, authors are required to send a printed copy of their paper, which exactly matches the copy approved for publication, along with any graphics that will appear in the paper.

TeX files may be submitted by email, FTP, or on diskette. The DVI file(s) and PostScript files should be submitted only by FTP or on diskette unless they are encoded properly to submit through email. (DVI files are binary and PostScript files tend to be very large.)

Electronically prepared manuscripts can be sent via email to `pub-submit@ams.org` (Internet). The subject line of the message should include the publication code to identify it as a Memoir. TeX source files, DVI files, and PostScript files can be transferred over the Internet by FTP to the Internet node `e-math.ams.org` (130.44.1.100).

Electronic graphics. Comprehensive instructions on preparing graphics are available at `www.ams.org/jourhtml/graphics.html`. A few of the major requirements are given here.

Submit files for graphics as EPS (Encapsulated PostScript) files. This includes graphics originated via a graphics application as well as scanned photographs or other computer-generated images. If this is not possible, TIFF files are acceptable as long as they can be opened in Adobe Photoshop or Illustrator. No matter what method was used to produce the graphic, it is necessary to provide a paper copy to the AMS.

Authors using graphics packages for the creation of electronic art should also avoid the use of any lines thinner than 0.5 points in width. Many graphics packages allow the user to specify a "hairline" for a very thin line. Hairlines often look acceptable when proofed on a typical laser printer. However, when produced on a high-resolution laser imagesetter, hairlines become nearly invisible and will be lost entirely in the final printing process.

Screens should be set to values between 15% and 85%. Screens which fall outside of this range are too light or too dark to print correctly. Variations of screens within a graphic should be no less than 10%.

Inquiries. Any inquiries concerning a paper that has been accepted for publication should be sent directly to the Electronic Prepress Department, American Mathematical Society, P. O. Box 6248, Providence, RI 02940-6248.

Editors

This journal is designed particularly for long research papers, normally at least 80 pages in length, and groups of cognate papers in pure and applied mathematics. Papers intended for publication in the *Memoirs* should be addressed to one of the following editors. In principle the Memoirs welcomes electronic submissions, and some of the editors, those whose names appear below with an asterisk (*), have indicated that they prefer them. However, editors reserve the right to request hard copies after papers have been submitted electronically. Authors are advised to make preliminary email inquiries to editors about whether they are likely to be able to handle submissions in a particular electronic form.

Selected Titles in This Series

(Continued from the front of this publication)

703 **Erik Guentner, Nigel Higson, and Jody Trout,** Equivariant E-theory for C^*-algebras, 2000

702 **Ilijas Farah,** Analytic guotients: Theory of liftings for quotients over analytic ideals on the integers, 2000

701 **Paul Selick and Jie Wu,** On natural coalgebra decompositions of tensor algebras and loop suspensions, 2000

700 **Vicente Cortés,** A new construction of homogeneous quaternionic manifolds and related geometric structures, 2000

699 **Alexander Fel′shtyn,** Dynamical zeta functions, Nielsen theory and Reidemeister torsion, 2000

698 **Andrew R. Kustin,** Complexes associated to two vectors and a rectangular matrix, 2000

697 **Deguang Han and David R. Larson,** Frames, bases and group representations, 2000

696 **Donald J. Estep, Mats G. Larson, and Roy D. Williams,** Estimating the error of numerical solutions of systems of reaction-diffusion equations, 2000

695 **Vitaly Bergelson and Randall McCutcheon,** An ergodic IP polynomial Szemerédi theorem, 2000

694 **Alberto Bressan, Graziano Crasta, and Benedetto Piccoli,** Well-posedness of the Cauchy problem for $n \times n$ systems of conservation laws, 2000

693 **Doug Pickrell,** Invariant measures for unitary groups associated to Kac-Moody Lie algebras, 2000

692 **Mara D. Neusel,** Inverse invariant theory and Steenrod operations, 2000

691 **Bruce Hughes and Stratos Prassidis,** Control and relaxation over the circle, 2000

690 **Robert Rumely, Chi Fong Lau, and Robert Varley,** Existence of the sectional capacity, 2000

689 **M. A. Dickmann and F. Miraglia,** Special groups: Boolean-theoretic methods in the theory of quadratic forms, 2000

688 **Piotr Hajłasz and Pekka Koskela,** Sobolev met Poincaré, 2000

687 **Guy David and Stephen Semmes,** Uniform rectifiability and quasiminimizing sets of arbitrary codimension, 2000

686 **L. Gaunce Lewis, Jr.,** Splitting theorems for certain equivariant spectra, 2000

685 **Jean-Luc Joly, Guy Metivier, and Jeffrey Rauch,** Caustics for dissipative semilinear oscillations, 2000

684 **Harvey I. Blau, Bangteng Xu, Z. Arad, E. Fisman, V. Miloslavsky, and M. Muzychuk,** Homogeneous integral table algebras of degree three: A trilogy, 2000

683 **Serge Bouc,** Non-additive exact functors and tensor induction for Mackey functors, 2000

682 **Martin Majewski,** ational homotopical models and uniqueness, 2000

681 **David P. Blecher, Paul S. Muhly, and Vern I. Paulsen,** Categories of operator modules (Morita equivalence and projective modules, 2000

680 **Joachim Zacharias,** Continuous tensor products and Arveson's spectral C^*-algebras, 2000

679 **Y. A. Abramovich and A. K. Kitover,** Inverses of disjointness preserving operators, 2000

678 **Wilhelm Stannat,** The theory of generalized Dirichlet forms and its applications in analysis and stochastics, 1999

677 **Volodymyr V. Lyubashenko,** Squared Hopf algebras, 1999

For a complete list of titles in this series, visit the AMS Bookstore at **www.ams.org/bookstore/**.